Alexander Weiß

Applications of Microscopic Modelling in Finance

Alexander Weiß

Applications of Microscopic Modelling in Finance

Stability and liquidity of financial markets exemplified by an agent-based model

Südwestdeutscher Verlag für Hochschulschriften

Impressum/Imprint (nur für Deutschland/ only for Germany)
Bibliografische Information der Deutschen Nationalbibliothek: Die Deutsche Nationalbibliothek verzeichnet diese Publikation in der Deutschen Nationalbibliografie; detaillierte bibliografische Daten sind im Internet über http://dnb.d-nb.de abrufbar.

Alle in diesem Buch genannten Marken und Produktnamen unterliegen warenzeichen-, marken- oder patentrechtlichem Schutz bzw. sind Warenzeichen oder eingetragene Warenzeichen der jeweiligen Inhaber. Die Wiedergabe von Marken, Produktnamen, Gebrauchsnamen, Handelsnamen, Warenbezeichnungen u.s.w. in diesem Werk berechtigt auch ohne besondere Kennzeichnung nicht zu der Annahme, dass solche Namen im Sinne der Warenzeichen- und Markenschutzgesetzgebung als frei zu betrachten wären und daher von jedermann benutzt werden dürften.

Verlag: Südwestdeutscher Verlag für Hochschulschriften Aktiengesellschaft & Co. KG
Dudweiler Landstr. 99, 66123 Saarbrücken, Deutschland
Telefon +49 681 37 20 271-1, Telefax +49 681 37 20 271-0
Email: info@svh-verlag.de
Zugl.: Berlin, TU, Diss., 2009

Herstellung in Deutschland:
Schaltungsdienst Lange o.H.G., Berlin
Books on Demand GmbH, Norderstedt
Reha GmbH, Saarbrücken
Amazon Distribution GmbH, Leipzig
ISBN: 978-3-8381-1351-7

Imprint (only for USA, GB)
Bibliographic information published by the Deutsche Nationalbibliothek: The Deutsche Nationalbibliothek lists this publication in the Deutsche Nationalbibliografie; detailed bibliographic data are available in the Internet at http://dnb.d-nb.de.

Any brand names and product names mentioned in this book are subject to trademark, brand or patent protection and are trademarks or registered trademarks of their respective holders. The use of brand names, product names, common names, trade names, product descriptions etc. even without a particular marking in this works is in no way to be construed to mean that such names may be regarded as unrestricted in respect of trademark and brand protection legislation and could thus be used by anyone.

Publisher: Südwestdeutscher Verlag für Hochschulschriften Aktiengesellschaft & Co. KG
Dudweiler Landstr. 99, 66123 Saarbrücken, Germany
Phone +49 681 37 20 271-1, Fax +49 681 37 20 271-0
Email: info@svh-verlag.de

Printed in the U.S.A.
Printed in the U.K. by (see last page)
ISBN: 978-3-8381-1351-7

Copyright © 2010 by the author and Südwestdeutscher Verlag für Hochschulschriften Aktiengesellschaft & Co. KG and licensors
All rights reserved. Saarbrücken 2010

Für meine Eltern

Preface

This book is based on my same-titled dissertation, which I submitted to the Technical University Berlin in August 2009. In the time after submission, I prepared for the oral examination and defended my work successfully in October. The defence became a fruitful discussion with the referees, establishing new perspectives on my results that I had not focused on before.

These new impulses are included in this book, and, furthermore, typos and ambiguities that had survived several proof readings before have been removed. Overall, the additions and corrections have led to that improved version of my thesis the dear reader is now holding in his or her hands.

Alexander Weiß

Berlin, February 2010

Contents

Preface   iii

1 Introduction   1
    1.1 Microscopic models of financial markets - a motivation   1
    1.2 Review of the results   5
       1.2.1 Stability of markets: The stalking particles   5
       1.2.2 Illiquidity in markets: The execution of large orders   7
    1.3 Outline of the chapters   8
    1.4 Acknowledgement   9

2 The Opinion Game   13
    2.1 The model   13
    2.2 An implementation   15
    2.3 Stylized Facts   17
       2.3.1 The absence of autocorrelations   18
       2.3.2 Volatility clustering   19

3 The stalking particles   21
    3.1 Introduction   21
    3.2 The construction of the particle system   28
       3.2.1 A deterministic approach...   29
       3.2.2 ...applied to random paths   34
    3.3 Stability of the Brownian particle system   35
       3.3.1 The main result for the Brownian motion   35
       3.3.2 Proof of Theorem 3.2.(i): the transient case   36

	3.3.3	Proof of Theorem 3.2.(ii): the recurrent case	45
3.4		An analogy for heavy-tailed Lévy processes	51
	3.4.1	$\mathfrak{C}(\alpha)$ - A class of Lévy processes	51
	3.4.2	The main result for heavy-tailed Lévy processes	54
	3.4.3	Preliminaries for both parts of the proof	54
	3.4.4	Proof of Theorem 3.10.(i): the transient case	56
	3.4.5	Proof of Theorem 3.10.(ii): the recurrent case	71

4 Optimal execution strategies for large orders — 79

4.1		Introduction	79
4.2		The AFS market model, Version 1 & 2	83
4.3		The Opinion Game extended to large orders	88
4.4		GAFS - a generalisation of the AFS model	91
	4.4.1	The shape of the Opinion Game's order book	92
	4.4.2	From determining ρ to the GAFS model, Version 2	93
	4.4.3	From determining ρ to the GAFS model, Version 1	100
4.5		The (G)AFS optimal strategies applied to the Opinion Game	104
	4.5.1	Results for Version 2	104
	4.5.2	Results for Version 1	106
4.6		Missing features of the (G)AFS model	107
4.7		Proofs of the Theorems 4.3 and 4.4	111
	4.7.1	The optimal strategy for Version 1	115
	4.7.2	The optimal strategy for Version 2	119

A SimStocki - a simulation tool for the Opinion Game — 125

A.1		An exterior view	126
A.2		An interior view	128
	A.2.1	`Simulation.java`	129
	A.2.2	`Calculation.java`	136
	A.2.3	`Trader.java`	144

B Index of symbols — 151

Chapter 1

Introduction

In this book, we deal with two aspects of financial markets:

- their stability depending on long-term investors, and
- the problem of illiquidity when large volumes are traded.

These issues have been motivated by observations in the *Opinion Game*, a microscopic model of a stock market by Bovier, Černý and Hryniv, and our results contribute to a deeper understanding of this model. Nevertheless, the results can also be understood in a model-independent, economic sense. In this work, we examine both ways of interpretation. To start with, we give a motivation for agent-based[a] modelling of financial markets followed by a discussion of the unique characteristics of the Opinion Game in comparison to other models. Afterwards, we summarise our main results. An outline of the remaining chapters and the acknowledgement conclude the introduction.

1.1 Microscopic models of financial markets - a motivation

Die Ursache liegt in der Zukunft.

Joseph Beuys

One of the main issues of financial mathematics is the pricing of *derivatives*, financial instruments whose prices depend on variables of *underlyings*. Common

---
[a]We use *agent-based modelling* and *microscopic modelling* interchangeably.

examples of underlying assets are shares, bonds, commodities and foreign currencies[b]; a customary variable for many underlyings is the price. In the following, we mainly think of derivatives with underlying shares of stock for two reasons: First, these underlyings are usually traded on organised markets, and their prices are the result of supply and demand. Second, they are conceptually simpler than a bond for example, as a bond price depends on the time to maturity, and has therefore an additional variable.

A common type of derivatives is an *option*. This is a contract to buy or sell the underlying for the fixed *strike price* at some time before or at the option's *date of expiry*. There are also other derivatives, such as *futures*, *forwards*, *swaps* and more sophisticated combinations of these[c], possibly with various underlyings, but their crucial feature can already be seen using the example of an option: the purchase of a derivative is a bet on the future evolution of the underlying's price. Only if it develops in the *right* direction, the owner profits from the purchase of the derivative. A *fair* purchase price depends therefore on the future market price of the underlying. Yet, since this price is unknown in the present, *realistic* assumptions on the underlyings' price evolution are crucial for the pricing of the derivative.

Already Bachelier (1900) was confronted with this problem in his fundamental *Théorie de la spéculation* where he considered financial instruments similar to futures and options with a bond of the French government as underlying (see Section 3.2 of Voit (2005) for a nice summary). He argued that the relative price fluctuations on a financial market resulted from many small, random trading actions of market participants, and he concluded that the underlying's price evolution should effectively behave like a geometric Brownian motion. Later research on this topic, including the famous work of Black and Scholes (1973) and Merton (1973) on option pricing for shares of common stock, has often picked up on this choice. Even today, extensions of this model play an important role in the pricing of derivatives (good text book references are Karatzas and Shreve (2001), and

---

[b]These examples are in accord with Baxter and Rennie (1998). Yet, there are more exotic underlyings such as the weather (see Dischel (2002)).

[c]We refer to Hull (2000) for a comprehensive overview.

Föllmer and Schied (2004)).

Nevertheless, there are some drawbacks in this ansatz. In the last decades, the increasing computing power have made it possible to analyse large quantities of financial data. These empirical studies have revealed statistical features of the price evolutions that consistently appear on different markets and on different time scales (see Cont (2001) for an overview). They are called *stylized facts*, and some of them contradict the assumption of a Brownian price evolution. For example, time intervals on which the price process has large fluctuations alternate with those intervals on which the fluctuations are small, a sign for a nontrivial correlation structure of the relative price changes.

Another issue is the assumption that the market price evolution is independent from the actions of the derivative's owner. The owner is a market participant such that he or she can influence the price by trading. This may be a minor problem with small investors, but it becomes important if the trader has enough financial power to trade volumes that are significant in comparison to the daily traded volume of the particular market (see Section 4.1 for references). Of course, the interaction effects between *large traders* and the price evolution are not restricted to the assumption that the price evolves like a Brownian motion.

These problems require a deeper understanding on the fundamental mechanisms that drive financial markets. Bachelier had already argued on the level of market participants and their trading actions to motivate his assumptions, but only Stigler (1964) (to our knowledge) also proposed a model on this level to get information about the emerging price process.[d] Today, there is a variety of *microscopic market models* with different characteristics and features.

We distinguish two groups of microscopic models with different strategies. The models of the first group consider a certain number of traders or agents[e] that assume states from a usually small state space; typical states are *buy*, *sell* and *hold*. The traders change their state by deterministic or stochastic rules that

---

[d] In particular, Stigler used a simple *order book* model to challenge both the statements and the assumptions of a report by the U.S. Securities and Exchange Commission (SEC) about the benefits of public regulation of certain financial markets.
[e] We use *trader* and *agent* interchangeably.

may depend on the previous evolution of the system. The price of the traded asset is usually given by a function that maps the market imbalance to the price change. Examples for this group of models are the *Minority Game*, introduced by Challet and Zhang (1997) and extended in various ways (see Challet, Marsili and Zhang (2004)), or the *Santa Fe Artificial Stock Market* (see LeBaron (2002) and references therein).

The models of the second group focus more on the price formation process. Instead of a deterministic rule that links the market's imbalance with the price, the traders can place orders, that is they can quote prices for which they are willing to trade. The orders are collected in the *order book*, which is *cleared* when necessary: If a buying and a selling trader have placed orders at the same price, a trade takes place and the particular orders are removed from the order book. The price of the traded asset is determined from the highest price of all *buy orders*, the *(best) bid price*, and the lowest price of all *sell orders*, the *(best) ask price*, in the cleared order book. This algorithm of trading is called *double auctioning* and it is widely used at stock exchanges.

The particular double auctioning models differ in the way they determine which traders place orders at which price. In some models, which seem to go back to the work of Mendelson (1982), the orders randomly appear with certain rates in the order book. If a buy order arrives at the price of an already placed sell order (or vice versa), the order book is cleared. It is also possible that orders are randomly removed with certain rates. Recent examples for this class of models are the works of Daniels et al. (2003), Smith et al. (2003) or Cont et al. (2008). In a second class of models, the price range of the order book is fixed, and orders only arrive at the boundaries. Then, they perform random walks, which may be inhomogeneous in space and time, until they hit orders of the other type (buy/sell) or until they leave the order book at the boundaries. This class of models seems to arise from the work of Bak et al. (1997) and has been picked up by Eliezer and Kogan (1998), and Tang and Tian (1999) for example.

Nevertheless, all these double auctioning models have one characteristic in common: They try to model the market in terms of the order book. Yet, only a small fraction of all tradable shares can be found in order books at any given time. On the other hand, also traders without placed orders have an opinion about the *right* price to trade and can quickly make an appearance if they think that the market evolves to their advantage. Bovier et al. (2006) introduced only recently a microscopic market model, called the *Opinion Game*, that is based on a *generalised order book*. In the Opinion Game, every market participant has a *subjective* opinion about the right price to trade. The opinions are collected in the generalised order book, which is cleared by double auctioning as in the models before. In the original paper, the authors explained the algorithm of the Opinion Game and demonstrated by simulations that the generated price process shows nontrivial statistical features that are also known from real markets. In a subsequent article, Bovier and Černý (2007) analysed the large scale asymptotics of the generalised order book's shape in vicinity to the *best quotes*, that is the ask price and the bid price.

1.2 Review of the results

In this work, we continue the analysis of the Opinion Game with both analytic and numerical tools.

1.2.1 Stability of markets: The stalking particles

Numerical simulations in the Opinion Game have indicated a dichotomy in the long-term behaviour of the model as a function of the parameters: Either the empirical distribution of the generalised order book stays tight, that is the ask price and the bid price remain at a bounded distance, or the generalised order book diverges in the sense that buyers and sellers drift infinitely far apart. A key parameter whose change triggers the transition between these alternatives is the rate at which each trader's activity decreases with the distance of his or her opinion from the current price. Clearly, only the former situation corresponds to a functioning market. It appears therefore of great interest to obtain a sound understanding of the mechanisms leading to instability.

To analyse this phenomenon, we introduce a simplified, tractable model that captures the main characteristics of the Opinion Game with a diverging generalised order book. It consists of three elements only: the current logarithmic price (L_t), the buyers' opinion (X_t), and the sellers' opinion (Y_t). The model represents the idea that, in a diverging order book, the price is generated by a few traders while the majority of market participants is far off the price and acts slowly, since the rate of activity decreases as function of the distance to the price. In particular, inspired by the Opinion Game, we define L to a be random process and X and Y to be solutions of the differential equations

$$\frac{d}{dt}X_t = (K + L_t - X_t)^{-\gamma}, \qquad (1.2.1)$$

$$\frac{d}{dt}Y_t = -(K + Y_t - L_t)^{-\gamma} \qquad (1.2.2)$$

for all t with $Y_t \geq L_t \geq X_t$. The positive constant K avoids a singularity if the distance to the price becomes 0, the positive constant γ controls how fast the activity decays if the distance becomes larger. When the price reaches one of the trader groups, the particular group does not provide enough resistance to stop or even reverse the price movement. Instead, the traders are pushed further away by the price. We implement this behaviour mathematically by the assumption that the relative order of Y, L and X is always preserved: If L_t is not an element of $[X_{t-}, Y_{t-}]$, X_t or Y_t adopts the position of L_t.

This *three particle system* can also be considered as a model for the behaviour of long-term investors on a financial market. These market participants do not speculate on fast profit, thus their opinion about the *right* share value can differ strongly from the current price. Furthermore, they do not react to small price fluctuations, but update their opinions only slowly, considering longer time scales.

In Weiß (2009), we have proven that the stability of such a market in terms of the behaviour of $Y - X$ undergoes a phase transition in the parameter γ if L is a Brownian motion. Namely, $Y_t - X_t$ is recurrent if $\gamma < 1$ and transient if $\gamma > 1$.

Here, we extend our results. We prove first, that we can have both recurrence and transience in the Brownian case if $\gamma = 1$. The particular behaviour then depends on the volatility of the Brownian motion: If it is below a certain threshold, the system is recurrent, if it is above a second threshold, the system is transient. Unfortunately, the two thresholds are not equal such that we are not able solve the critical case completely.

We have already discussed that a Brownian motion is a suboptimal choice for a price process. Lévy processes are better suited (see Section 3.1 for a discussion and references). We prove that the particle system is recurrent for $\gamma < \alpha - 1$ and transient for $\gamma > \alpha - 1$ if L is an α-stable Lévy process with both positive and negative jumps. As the heavy tails are the only crucial property we need for the proof, we can even extend the result to a very general class of heavy-tailed Lévy processes.

1.2.2 Illiquidity in markets: The execution of large orders

One of the problems of great current interest for banks is to analyse the effect of large transactions on the market price of the traded asset with the aim of finding optimal strategies to execute such transactions with a minimum of negative monetary effects. The approaches to this issue (for references see Section 4.1) differ strongly in the assumptions on how the market responses to the transactions, resulting in different *optimal* strategies.

In this work, we test optimal trading strategies numerically with the Opinion Game. In particular, we focus on the strategies that are suggested by Alfonsi, Fruth and Schied (2010). In their model, the market is described by the *static* shape of the order book, and a constant governing the book's speed of recovery after a large order has been executed. Our simulations suggest that a constant does not describe the recovery process sufficiently; there are dependencies on the order size.

We thus generalise the market model such that the recovery speed of the order book depends on the size of the impact. For the generalisation, we then prove optimal trading strategies, and demonstrate by simulations that they perform better than the original strategies.

Despite our improvements, the market model still performs poorly in an absolute sense, that is the costs of a large order predicted by the model are not even approximately identical to the numerical results. We argue that, for the Opinion Game, these differences results from the absence of a *permanent impact* and the assumption of a static order book shape. Since both features, the existence of permanent impact and a dynamic order book shape, are known to exist on real financial markets, our observations suggest possible improvements for the model.

This part of the work has partially been published in Weiß (2010).

1.3 Outline of the chapters

The Opinion Game is the origin of our work; we summarise the results from Bovier et al. (2006) in Chapter 2. There, we focus on the general algorithm (Section 2.1), and the particular implementation that we have used for the numerical analyses (Section 2.2). We conclude the chapter with some statistical features of the Opinion Game (Section 2.3).

Chapter 3 is devoted to the three particle system. We motivate the model in more detail by numerical results from the Opinion Game in Section 3.1, and construct it formally (Section 3.2). Afterwards, we state and prove the main results for the Brownian case (Section 3.3) and the Lévy case (Section 3.4).

We deal with the execution of large orders in Chapter 4. We first introduce the market model of Alfonsi et al. (2010), and cite the main theorems of this paper about the optimal trading strategies (Section 4.2). Afterwards, we discuss how the algorithm of the Opinion Game can be extended to realise large orders (Section 4.3), and illustrate the problems to calibrate the model for the Opinion Game, which leads to the generalised market model and the particular optimal trading strategies (Section 4.4). In Section 4.5, we show that the generalised strategies are an improvement. Nevertheless, the theoretically predicted costs still deviate strongly from the empirical costs. The reasons for the deviations are topic of Section 4.6. The proofs for the optimal trading strategies of the generalised model can be found in Section 4.7.

The numerical results in this thesis would not have been possible without a powerful simulation tool. We have developed SIMSTOCKI as simulation platform for the Opinion Game. In Appendix A, we give an insight into the program. The functionality of SIMSTOCKI is topic of Section A.1, a discussion of the class structure and extracts of the source code can be found in Section A.2.

There are some notations in this work that we use without defining them, since we think that they belong to the conventional vocabulary of mathematics. Appendix B contains a summary of them and shortly explains their meanings.

Due to the finite number of available characters and symbols, we could not avoid using some of them twice to denote different objects. However, the spatial (measured in pages) and thematic distance between two objects with the same appellation is large enough to avoid obscurities. Let us furthermore remark that the word "Theorem" is printed in cursive characters for cited propositions to distinguish them clearly from our own results.

1.4 Acknowledgement

This acknowledgement is condemned to be incomplete. The impulses influencing my work in the last years were just too various and sometimes also too subtle. I can do nothing more than to restrict myself to the most important people (with respect to a rather subjective measure) and to apologise to all the others for not mentioning them here explicitly. My gratitude comes from the depth of my heart, thus I allow myself to continue in German, saving my thoughts the additional way into another language.

Mein größter Dank gilt meinem Betreuer Herrn Prof. Dr. Bovier, der mich auf so vielfältige Weise unterstützt hat und ohne den diese Arbeit nicht möglich gewesen wäre. Er gab mir die Möglichkeit, mein Dissertationsthema innerhalb seines Forschungsbereiches frei zu wählen, und war mir während meiner gesamten Promotionszeit ein interessierter und inspirierender Diskussionspartner, der mich einerseits meine eigenen Ideen ausprobieren ließ und andererseits darauf achtete, dass meine Gedanken wieder in die richtigen Bahnen gelenkt wurden, wenn sie auf allzu schlammige Pfade abgekommen waren. Neben seiner fachlichen Unterstützung, sorgte er für eine ausgezeichnete Arbeitsumgebung.

Diese Umgebung wurde zu wesentlichen Teilen von drei Institutionen getragen: dem Weierstraß-Institut für Angewandte Analysis und Stochastik (WIAS), dem DFG Forschungszentrum Matheon und der International Research Training Group "Stochastic Models of Complex Processes" (IRTG), denen ich für ihre mannigfaltige Unterstützung aus tiefsten Herzen dankbar bin.

Als Mitglied der Arbeitsgruppe "Stochastische Systeme mit Wechselwirkungen" am WIAS bot sich mir die Möglichkeit, zahlreiche exzellente Stochastiker zu treffen, die entweder Gäste oder auch Mitglieder der Arbeitsgruppe waren. Insbesondere die Diskussionen mit Prof. Dr. Matthias Birkner, Dr. Nicholas Champagnat, Dr. Jiří Černý und Martin Slowik haben mir immer wieder geholfen, meine Gedanken zu ordnen, und mich auf neue Ideen gebracht. Danke, merci und děkuji. Christina van de Sand danke ich für ihre zuverlässige Unterstützung in der Administration. Dem Institut selbst danke ich für den exzellenten Arbeitsplatz, der mir zur Verfügung stand.

Die großzügige finanzielle Unterstützung des Matheon ermöglichte es mir, mich ganz auf meine Promotion konzentrieren zu können. Die Veranstaltungen, die vom Forschungszentrum initiiert wurden, vertieften mein Verständnis für das Zusammenspiel verschiedener mathematischer Disziplinen zur Lösung industrieller Probleme. Insbesondere die Nachwuchswochenenden halfen, die grundlegenden Probleme anderer Teildisziplinen zu verstehen und die Berliner Mathematikergemeinschaft außerhalb der Stochastik kennenzulernen.

Wenn das WIAS meine physische Heimat während der Promotion war, so war die IRTG meine soziale. Aus den Kollegiaten und Stipendiaten wurden Freunde, die obligatorischen Vorträge im IRTG-Seminar verbesserten meine *Soft Skills* und die vielfältigen fachlichen Veranstaltungen vertieften mein stochastisches Wissen. Insbesondere möchte ich mich bei den drei PostDocs Dr. Evangelia Petrou, Dr.

Noemi Kurt und Dr. Yvon Vignaud bedanken, die viel Mühe und Zeit in die Organisation der IRTG gesteckt haben und dennoch bei Fragen und Problemen immer zur Verfügung standen.

Mein Dank gilt weiterhin einigen Personen, die sehr konkret an der Entstehung der Dissertationsschrift mitgewirkt haben. Dr. Frank Aurzada hat maßgeblich zu Kapitel 3 beigetragen. Er hörte sich nicht nur geduldig meine mathematischen Probleme an, die auftraten, als ich das Drei-Teilchen-Modell auf Lévy-Prozesse erweitern wollte, sondern half mir auch mit zahlreichen Bemerkungen zu einer früheren Version der Dissertationsschrift dabei, den Aufbau des Kapitels besser zu strukturieren. Dr. Evangelia Petrou (ja, nochmal!) investierte unzählige Stunden in Diskussionen mit mir über den Artikel, der Kapitel 4 zugrunde liegt. Ohne ihr Verständnis, ihre Geduld und ihre Fähigkeit, ein Arbeitstreffen wie einen freundschaftlichen Plausch wirken zu lassen, wäre ich um viel Frust reicher und diese Arbeit um ein Kapitel ärmer. ευχαριστώ. Antje Fruth danke ich für ihre Hilfsbereitschaft bei meinen Fragen zum Originalmodell, das ich in Kapitel 4 verallgemeinere. Erst durch die Prüfungskommission ist diese Arbeit zu einer Doktorarbeit geworden. Ich möchte neben meinem Betreuer dem zweiten Gutachter Prof. Dr. Peter Bank und dem Vorsitzenden der Kommission Prof. Dr. Martin Skutella für ihr Engagement danken.

Meinen Eltern kann ich nicht genug danken für ihre immerwährende Unterstützung und ihr Vertrauen in mich. Ihr seht, man kann durchaus mathematisch forschen, auch wenn schon alle Zahlen bekannt sind. Linda Guttowski danke ich für die vorgeschlagenen Korrekturen in der Arbeit (insofern gehört ihr Name auch in den vorigen Absatz), aber noch viel mehr (und deshalb steht er hier) für ihre Kraft, Zuneigung und das Kraulen. Jetzt bist Du dran mit Siegen!

Chapter 2

The Opinion Game

The Opinion Game as introduced by Bovier et al. (2006) describes a whole class of agent-based models for financial markets. In the particular article, it is first sketched as *rough* algorithm that mainly captures the idea to simulate a market by considering all market participants with their subjective opinions about the right trading price. A specific implementation follows only afterwards. We adopt this structure. In the next section, we state the general algorithm for the Opinion Game, and substantiate this *skeleton* in Section 2.2. The emerging price process of the implementation has some interesting correlation features, as we show in Section 2.3.

2.1 The model

The Opinion Game considers a market with N traders trading $M < N$ shares, $M, N \in \mathbb{N}$. For simplification, every trader can own at most one share; furthermore, a discrete time and space setting is assumed. The discreteness assumptions are inessential but simplify computer simulations. The state of trader i is given by his or her opinion about the share value[a] $p_i \in \epsilon\mathbb{Z}$, and the number of possessed stocks $n_i \in \{0, 1\}$. The parameter $\epsilon > 0$ is called the *tick size* of the price scale. A trader with a share is called a *seller*, one without a share is called a *buyer*. The state of the generalised order book[b] is given by the entirety of all traders' states.

---

[a]In this chapter, we strictly distinguish between the *price* and the *value* of a share. While the price is *objectively* determined from the market situation, the value denotes the *subjective right* trading price for every trader.

[b]If no confusion is caused, we neglect the word "generalised".

A state is said to be *stable* if the traders with the M highest opinions possess shares. In particular, a stable order book state can be completely described by the traders' opinions $\mathbf{p} := (p_1, \ldots, p_N)$. For stable states, one can define the *(best) ask price* as the lowest opinion of all traders possessing a share:

$$p^a := \min\{p_i : n_i = 1\}; \qquad (2.1.1)$$

the *(best) bid price* is defined as the highest opinion of all traders without a share:

$$p^b := \max\{p_i : n_i = 0\}. \qquad (2.1.2)$$

The *current logarithmic*[c] *price* of the stock is given by $p := (p^a + p^b)/2$. The updating of the order book's state, $\mathbf{p}$, happens in three steps:

1. At time $(t+1) \in \mathbb{N}$, trader i is selected with probability

$$g(i; \mathbf{p}(t), t). \qquad (2.1.3)$$

2. The selected trader i changes his or her opinion to $p_i(t) + d$ with $d \in \epsilon\mathbb{Z}$ having distribution $f(\cdot; \mathbf{p}(t), i, t)$.

3. If $\mathbf{p}' = (p_1(t), \ldots, p_i(t) + d, \cdots, p_N(t))$ is stable, then $\mathbf{p}(t+1) := \mathbf{p}'$. Otherwise, trader i exchanges the state of ownership, $n_i(t)$, with the lowest asker, respectively highest bidder, j. If there are several agents at the best price, one is uniformly chosen. Afterwards, to avoid a direct re-trade, both participants change their opinion away from the trading price.

There may be some irritating assumptions that must be argued. First, every trader can only possess one share at most. At this point, one may prefer not to think of single traders, but consider single shares or demands instead. These carry labels with their values that are allocated to them by their owners. Since real traders may have strategies involving sales of various amounts of shares for

---
[c]"*In the whole of modern financial literature, it is postulated that the relevant variable to study is not the [absolute] increment [...] itself, but rather [the relative price change][...].*" (Bouchaud and Potters (2005), Section 6.2.1) Consequently, the logarithmic price is the interesting quantity to study. Since we consider that the whole system exists on a logarithmic scale, we neglect the word "logarithmic", like we already did it when defining the traders' opinions and the best prices.

different prices at the same time, this interpretation is not unrealistic.

Second, **p** is a Markov chain. This may seem strange, since price charts are easily accessible information, which are taken at least partly into account by serious traders. Yet, it is difficult to find reasonable update rules for the traders' opinions that are based on the price history. Furthermore, the charts are mostly analysed to get information about the current opinions of the other traders. One may therefore assume that, instead of considering the past, the traders have access to the prevailing current opinion (by rumours or newspapers) and consider this information.

Last, there is no specific concept of money contained in the Opinion Game. In particular, the traders' credit balances are not observed, and the objective trading successes of the single market participants are not taken into account. It seems reasonable to assume that no trader invests a substantial fraction of his or her money into the particular asset such that all traders are able to purchase shares when they want to, in case of need maybe by credit. It is also unclear how a bad trading performance would influence the particular trading strategy. What is important instead, and what is also incorporated in the Opinion Game, is the fact that any trader has the subjective impression to make profit at any trade. This implies that the subjective opinion must be wrong for at least one of the traders, but it seems to reflect reality.[d]

By now, we have introduced a general model whose main feature is the description of a share price's time evolution as the result of a random interaction process that reflects the change of the individual traders' opinions on the value of the share. Observe that **p** is Markovian, but the resulting price is usually not.

2.2 An implementation

The specific implementation for all that is following makes four reasonable assumptions:

---

[d] *"I agree with no man's opinion. [...] I have some of my own."* - Ivan Sergeyevich Turgenev, *Fathers and Sons*, Chapter XIII.

- The traders update their opinions in dependence on the current best ask and best bid price, and have a tendency to move into the direction of the best prices.

- A financial market is not a closed system but interacts with the rest of the world. External events, for example natural disasters, political conflicts or rumours, influence the traders.

- Traders with opinions far away from the current price update their opinions less frequently than traders in vicinity of the best prices.

- After a trade, the participating traders update their opinion again, since they both have the impression they have made a *good deal*. The new owner of the traded share increases his or her opinion, expecting a higher trading price; the new buyer equally decreases his or her opinion.

These assumptions are fulfilled by the following implementation:

The function g is defined by

$$g(i; \mathbf{p}(t), t) := h(p_i(t) - p^\bullet(t))/Z_g(\mathbf{p}(t)) \tag{2.2.1}$$

with

$$h(x) := 1/(1 + |x|)^\gamma, \ \gamma > 0, \tag{2.2.2}$$

and Z_g normalising g such that $\sum_{i=1}^N g(i; \mathbf{p}(t), t) = 1$. If trader i is a buyer, $p^\bullet$ is the best bid price, p^b; else, it is the best ask price, p^a.

The size of d is chosen from the set $\{-\epsilon l, \dots, \epsilon l\}$, $l \in \mathbb{N}$, with probability

$$f(d; \mathbf{p}(t), i, t) := \frac{1}{2l+1} \left(\left(\delta_{p_i, p(t)} \delta_{\text{ext}}(t) \right)^d e^{V(p_i) - V(p_i + d)} \wedge 1 \right) \text{ for } d \neq 0, \tag{2.2.3}$$

and

$$f(0; \mathbf{p}(t), i, t) = 1 - \sum_{0 < |k| \leq l} f(k; \mathbf{p}(t), i, t). \tag{2.2.4}$$

The function $V : \mathbb{R} \to \mathbb{R}$ is an external potential. The parameter $\delta_{p_i, p(t)}$ describes the tendency to change the opinion into the direction of the price. Thus it is larger than 1 for $p_i < p$ and smaller for $p_i > p$. The second parameter, δ_{ext}, simulates the influences of the external events on the change of the opinions. This force is the same for all traders, but it can change its (usually random) strength in time. It is substantiated in the next section.

The jump away from the trading price in the last step is implemented by setting

$$p_i(t+1) = p^b(t) - g, \quad p_j(t+1) = p^b(t) + \bar{g} \qquad (2.2.5)$$

if trader i sells a share in this step, and

$$p_i(t+1) = p^a(t) + g, \quad p_j(t+1) = p^a(t) - \bar{g} \qquad (2.2.6)$$

if he or she buys it. Here, g and $\bar{g}$ are random numbers in $\epsilon\mathbb{N}$.

2.3 Stylized Facts

♯ of traders N	2000
♯ of shares M	1000
speed of adaptation γ	1.5
tick size ϵ	1
jump range $\{-l,\ldots,l\}$	$\{-4,\ldots,4\}$
drift of buyers $\delta_{p_i,p(t)}, p_i < p(t)$	$e^{0.1}$
drift of sellers $\delta_{p_i,p(t)}, p_i > p(t)$	$e^{-0.1}$
potential $V(x)$	1 for all x
jump ranges $g, \bar{g}$	random variables, uniformly distributed on $\{5,\ldots,20\}$, sampled independently every time they are used

Table 2.1: The standard parameters of the Opinion Game for our simulations.

Unless stated otherwise, we used the parameter values from Table 2.1 in our simulations. For the external influence, δ_{ext}, we realised a Poisson process with rate $1/2000$. When the Poisson process jumped for the ith time, we set δ_{ext} to $\exp(\epsilon_i s_i)$ with the random variables ϵ_i defined by $P(\epsilon_i = \pm 1) = 1/2$, and s_i being exponentially distributed with mean 0.12. All ϵ_i and s_i were mutually independent. Observe that, in expectation, the external force was slightly stronger than the drift to the price, and that the averaged length of a period with constant external drift strength was 2000 *simulation steps*. By simulation step, we mean here and in the following one run of the updating procedure, which includes choosing a trader, changing the opinion, trading if necessary.

We finish this chapter with an analysis of the correlation structure of the price process. We simulated the Opinion Game with the setting from Table 2.1 for 500 000 000 steps, and recorded the price all 100 steps. The *returns*, that is the relative price changes, are just the differences between the successive prices. We denoted the empirical returns by (η_k) with k ranging from 1 to 5 000 000 circa[e].

2.3.1 The absence of autocorrelations

The *temporal two-point correlation function* $C(l)$ is defined by[f]

$$C(l) := \frac{\langle \eta_k \eta_{k+l} \rangle}{\sigma^2}; \quad \sigma^2 := \langle \eta_k^2 \rangle, \tag{2.3.1}$$

where the notation $\langle \ldots \rangle$ refers to the empirical average over the whole time

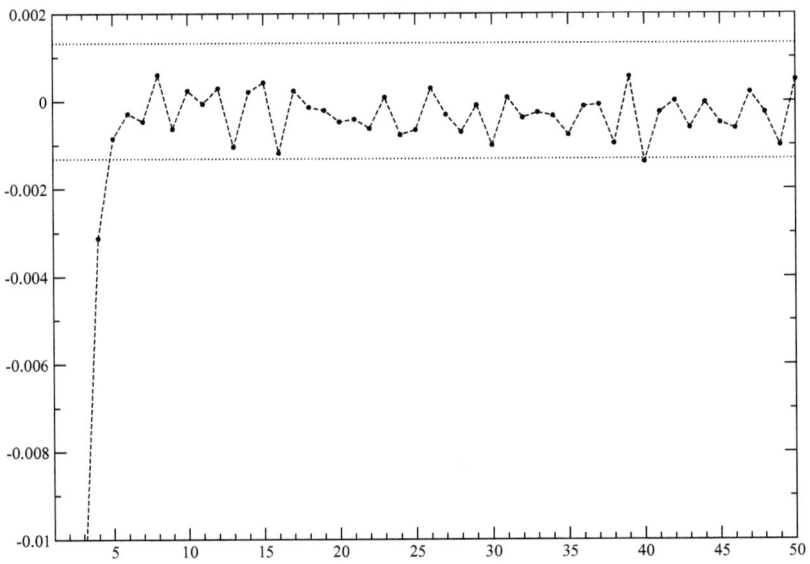

Figure 2.1: The graph of the mapping $l \mapsto C(l)$ with $C(l)$ being the autocorrelation function as defined in (2.3.1)

series. Bouchaud and Potters (2005) point out that, for uncorrelated increments, $C(l)$ should be equal to zero for $l \neq 0$ with an empirical standard deviation equal

---
[e]We gave the system about 2 000 000 steps to initialise.
[f]One should actually subtract the mean $\langle \eta \rangle$ from η. In our data series, however, we found $\langle \eta \rangle \approx 3.5 \cdot 10^{-5}$, which is negligible.

to $\sigma_e = 1/\sqrt{N}$, where N is the number of elements in the data set. Figure 2.1 shows the function $C(l)$ for the data of the Opinion Game. The dotted lines mark the significance interval $[-3\sigma_e, 3\sigma_e]$. While there is no statistically significant correlation of η_k and η_{k+l} for $l \geq 5$, we observe an anticorrelation for $l < 5$. The same behaviour is known for various assets, such as stock indices, exchange rates or interest rate indices (with significant correlations existing up to 30 minutes), and it is explained by the persistence of the order book: The orders in vicinity to the best quotes oppose *barriers* to the progression of the price (see Section 6.2.2 in Bouchaud and Potters (2005)). Recall that the average length of the external signal was 2000 simulation steps. On the other hand, we lose significant linear autocorrelations already after 500 steps. We may conclude that the trends induced by external signal does not dominate the market on these short time scales and with a focus on this kind of correlation.

2.3.2 Volatility clustering

We turn to long-ranged correlations, in particular, to *volatility clustering*. That is, in the words of Mandelbrot (1963),

> "[...] large changes tend to be followed by large changes - of either sign - and small changes tend to be followed by small changes [...]".

A way to measure these fluctuations is the *volatility variogram* $V(l)$ defined by

$$V(l) := \left\langle \left(\sigma_m^2 - \sigma_{m+l}^2\right)^2 \right\rangle. \tag{2.3.2}$$

The variables (σ_m^2) are given by

$$\sigma_m^2 := \frac{1}{100} \sum_{k=m}^{m+99} \eta_k^2 \tag{2.3.3}$$

and measure the empiric volatility of the returns. Figure 2.2 shows the volatility variogram for the time series of the Opinion Game. Bouchaud and Potters (2005)

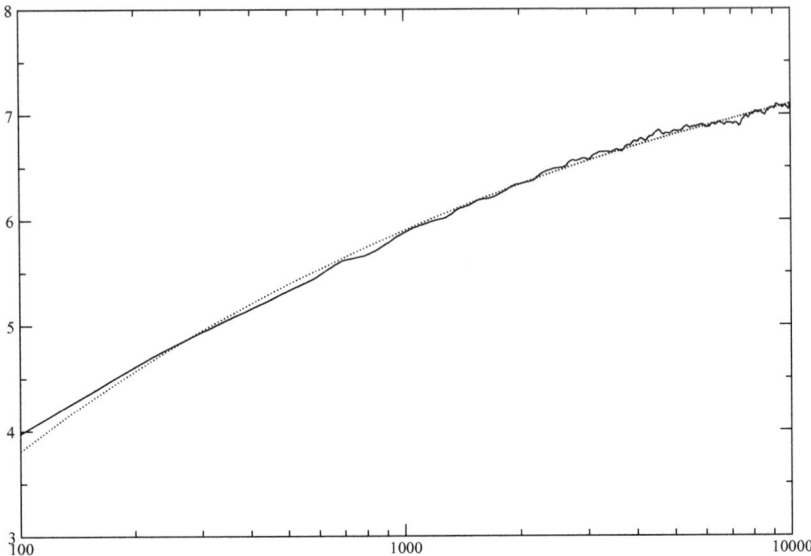

Figure 2.2: The graph of the mapping $l \mapsto V(l)$ (solid line) and the corresponding fitted power law (dotted line). Observe that we use a logarithmic scale for l.

argue that it can be well approximated by a power law of the form

$$\nu_\infty - \frac{\nu_0}{l^\nu}, \ l \neq 0, \tag{2.3.4}$$

ν_0, ν_∞ and ν being three degrees of freedom. An analysis of real markets finds $\nu \approx 0.22$ for individual stocks (Section 7.2.2). Applying a Newton-Gauß algorithm to the Opinion Game's data results in $\nu \approx 0.24$. Observe that the power law holds up to $l = 10^4$ for the Opinion Game, which is 500 times longer than the average length of the external signal; a memory effect that we explain in the next chapter.

Chapter 3

The stalking particles

3.1 Introduction

We have seen in Section 2.3 that the price process generated by the dynamics of the Opinion Game has some nontrivial correlations: While a significant linear autocorrelation of the returns can only be demonstrated for short time horizons, there is volatility clustering on time scales much longer than the parameter values give reason to expect. Two features of the implementation mainly cause this behaviour: the external signal δ_{ext} and the trader's decreasing updating speed as a function of the distance to the current price.

The external, time-inhomogeneous signal brings excitement into the market. As the traders basically perform random walks with drift to the price, an Opinion Game market without that signal shows lower, more uniform volatility.

The decaying updating speed of the agents gives memory to the market. Suppose the external signal generates a strong, positive drift, inducing a preference of the agents to move up and, consequently, an upward price movement. This shift does not work symmetric for buyers and sellers. The price moves into the direction of the sellers, causing a higher updating rate. Additionally, the sellers have a tendency to move to the ask price that impedes the upward movement. The group of sellers is squeezed by these two opposed drifts. On the other hand, only some buyers are able to stay close to the price. The remaining ones have not been able to follow the sudden price movement in the beginning; now, their distance to the price is already too large to catch up. The group of buyers is stretched.

In this situation, we can qualitatively deduce the direction and strength of the external drift from the current static order book configuration only.

Let us now suppose that the external signal changes its sign such that there is a drift downwards. In principle, this change just reverses the roles of buyers and sellers. However, the imbalanced geometries of the groups intensify the effect of the external signal. The stretched group provides less resistance against the price movement in its direction, since the number of buyers at the single prices is low. As a result, the gap between buyers and sellers grows. The externally induced drift may eventually become small again, but, in the meantime, the price, driven by a few traders, fluctuates within the enlarged *gap* without much resistance. These fluctuations cause a temporary higher volatility, and, on a larger time scale, volatility clustering.

The dependence of the traders' updating speed on their distance to the current price is consequently of paramount importance. In particular, recall from Section 2.2 that the probability of agent i with current opinion p_i to be chosen is given by

$$\frac{(1+|p_i - p^\bullet|)^{-\gamma}}{\sum_{j=0}^{N}(1+|p_j - p^\bullet|)^{-\gamma}}, \gamma > 0, \tag{3.1.1}$$

a summary of the lines (2.1.3) and (2.2.2). Simulations show that a larger value of γ encourages a higher volatility. Yet, a larger γ also increases the risk of a market instability. If the gap size exceeds a γ-dependent threshold, the market cannot recover, that is the gap size decreases no more. Instead, two agents, a buyer and a seller, determine the price by trading with each other within the gap. Their direction is mainly driven by the external signal. All other traders can hardly move as their distance to the current price is huge in comparison to the two agents in the gap. A strong external drift may sometimes push the two traders in proximity to one of the trading groups, but then, the same drift pushes that group further away from the other market participants. If the drift changes again, the decay in the updating speed is so steep that only one member is able to escape from the group.

In simulations, unstable market situations occur quite fast for $\gamma \geq 1.6$, yet,

3.1. INTRODUCTION

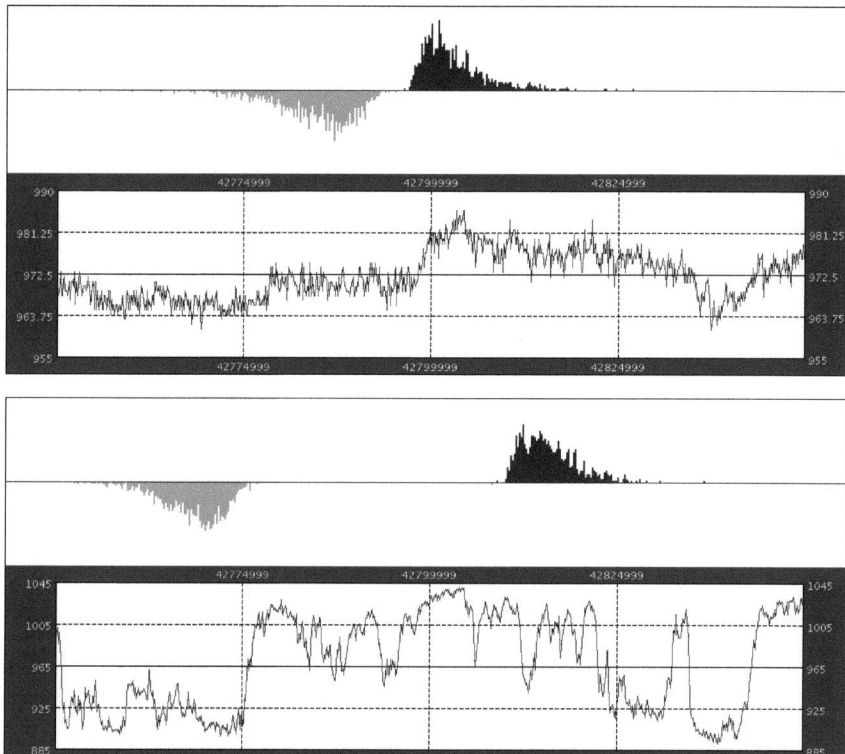

Figure 3.1: Screenshots of the virtual order books after 428 500 simulation steps for $\gamma = 1.5$ (top) and $\gamma = 1.6$ (bottom) with the same initial conditions and the same realisation of external influences. Observe the different distances between buyers (light grey) and sellers (dark grey) and the different behaviour of the price processes in the box below.

the model seems to remain stable for $\gamma = 1.5$. On the other hand, if we start a simulation with $\gamma = 1.5$ and a large gap, the system does not recover. Figures 3.1 to 3.3 illustrate these statements. For Figures 3.2 and 3.3, we recorded the distance between the 950th and the 1050th trader ordered by their opinions (in other words, the buyer with the 50th highest opinion and the seller with the 50th lowest one). The recorded quantity is not the gap in the strict sense, which is the difference between ask and bid price, but our choice is less volatile such that the qualitative behaviour is better observable.

Our observations suggest the question if there is a critical value for γ such that the market remains stable if γ is chosen smaller than that value, or if there is a

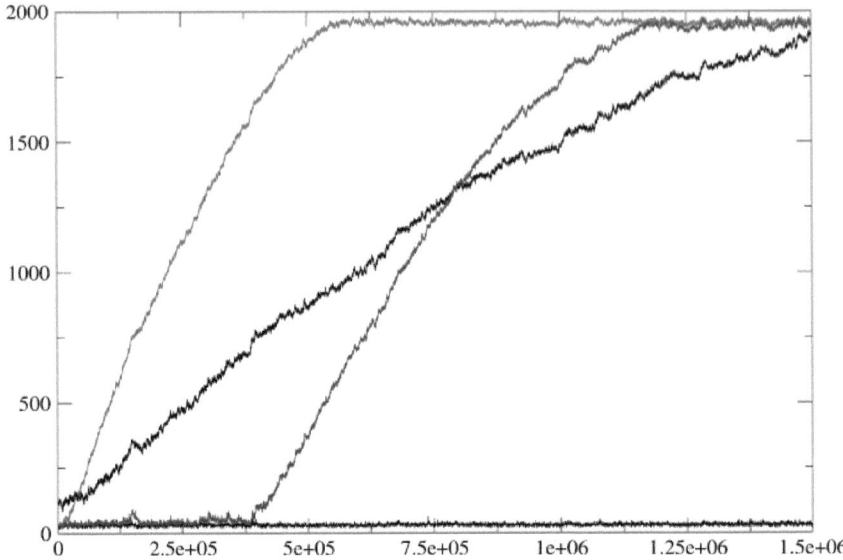

Figure 3.2: The gap of the system for different γ. Being stable for $\gamma = 1.5$ (black lower graph), it increases for $\gamma = 1.6$ (dark grey) and $\gamma = 1.7$ (light grey). If the system is started with $\gamma = 1.5$ but with an artificially enlarged gap, it also increases (black increasing graph). The convergence to a value below 2000 is due to a restriction of the state space in the simulations.

γ-depending, finite threshold for all positive values of γ such that the gap cannot recover anymore, once its size is larger than that threshold. For an answer, we consider a simplified model that still captures the main features of the Opinion Game in an unstable market situation.

Let us observe an unstable market situation in detail. $N-2$ agents are members of the two squeezed trading groups, which have a large distance to each other. The remaining two traders, in proximity to each other, are located in between; for the moment, let us assume that they are close to the middle of the gap. Then, the probability to be chosen is roughly the same for every member of the two groups: It is of order $(1 + \text{gap}/2)^{-\gamma}$ with "gap" denoting the gap size. The two remaining agents are mainly driven by the external drift, which is random in direction, strength and time. Neglecting the particular effects of the two agents *hitting* a group, we simply assume that the whole group cannot resist the drift in this situation and follows it until it reverses its direction. These assumptions lead to the three particle model. Two particles represent the two large groups,

3.1. INTRODUCTION

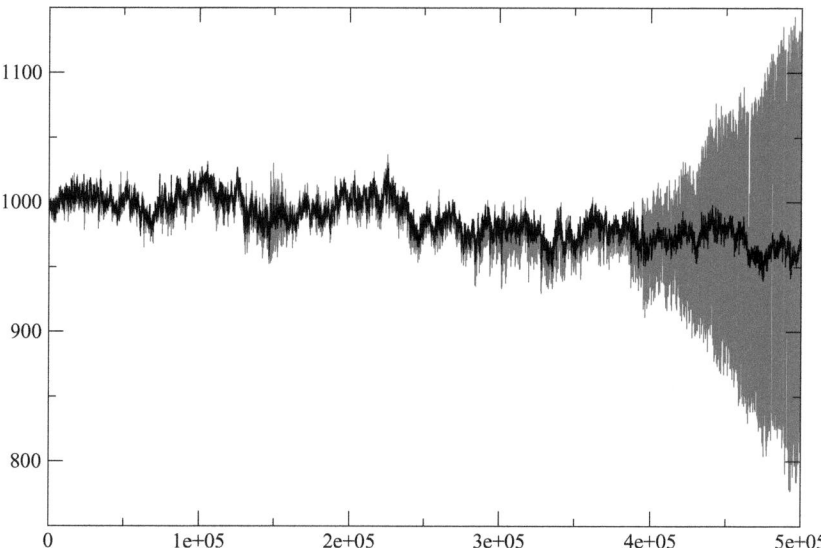

Figure 3.3: The stable, respectively unstable, behaviour for $\gamma = 1.5$ (black) and $\gamma = 1.6$ (grey) in terms of the price process.

the third particle stands for the two traders determining the price or, simpler, for the price[a] itself. We denote the positions of the three particles by X (the buyer group), Y (the seller group) and L (the price). The dynamics of X and Y are given by

$$\frac{d}{dt} X_t = (K + L_t - X_t)^{-\gamma}, \ \gamma > 0 \qquad (3.1.2)$$

and

$$\frac{d}{dt} Y_t = -(K + Y_t - L_t)^{-\gamma}, \qquad (3.1.3)$$

as long as $X_t < L_t < Y_t$. We slightly generalise the model by considering an arbitrary constant $K > 0$. L is a random process that we specify in a moment. We would like to preserve the relative order of the three particles to each other for all times. Thus, if L_t leaves the interval $[X_{t-}, Y_{t-}]$, we require that the system immediately recovers the order by setting X_t or Y_t, depending at which boundary L has left the interval, at the same position as L_t. The formal construction of the three particle system is the issue of Section 3.2.

---
[a] Let us stress again that the price axis of the generalised order book is logarithmic. Thus, we always mean the logarithmic price when talking about the price.

We can also interpret the three particle model as a simple example for the behaviour of long-term investors, which do not trade on the small price fluctuations, but expect large price changes. This expectation is reflected in the investors' trading strategies. Their opinions concerning the right trading price are far off the current market price and change only slowly; the small, *daily* price fluctuations are basically ignored. Periodically, the price matches the investors' opinions and shares are bought and sold. Afterwards, the individuals within the group of long-term investors may have changed, but the group itself persists.

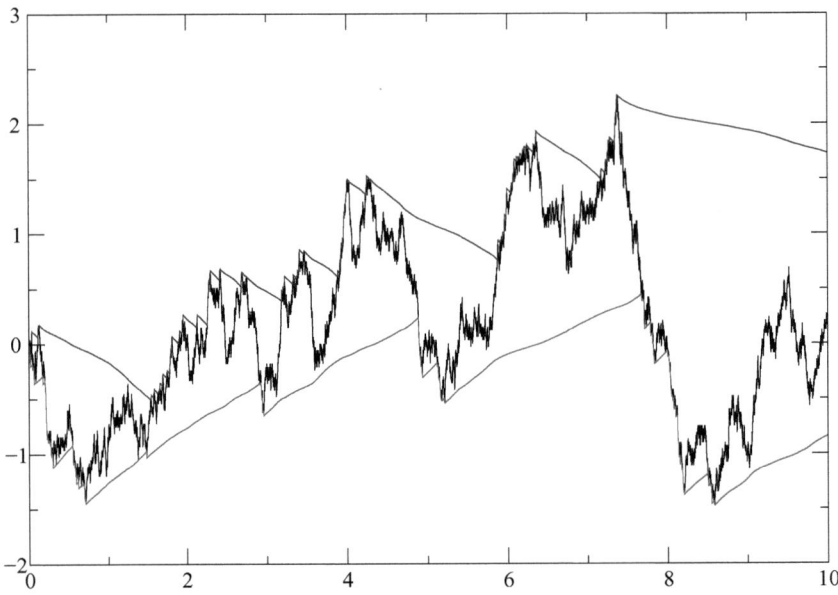

Figure 3.4: The price L (black), and the opinions X (light grey, bottom) and Y (dark grey, top) evolving in time. Here, L is chosen to be a Brownian motion.

In Weiß (2009), we analysed the three particle model for L being a Brownian motion (see Figure 3.4), and found that $\gamma = 1$ is the critical value for the system's stability. In particular, the last time that the distance between X and Y is smaller than r is infinite for $\gamma < 1$ and finite for $\gamma > 1$. The statement holds almost surely for all $r > 0$. There is a heuristic argument why 1 is the critical value. For a constant $c > 0$, we scale time by c^2 and space by c. We denote the processes' scaled versions by adding a superscripted c. By the scaling property of a Brownian

3.1. INTRODUCTION

motion, we have that L^c is equal to L in distribution. On the other hand, X^c solves the scaled differential equation

$$\frac{d}{dt}X_t^c = \frac{c^{1-\gamma}}{(K/c + L_t^c - X_t^c)^\gamma}. \tag{3.1.4}$$

If one assumes $L_t^c - X_t^c$ to be positive, the slope becomes infinite for $\gamma < 1$ and 0 for $\gamma > 1$ as c tends to infinity. This observation suggests that $Y - X$ remains stable for $\gamma < 1$ only.

We extend this result in the following sections. First, we refine the proof of Weiß (2009) and get a result for the critical case, $\gamma = 1$. In this case, the question of recurrence and transience depends on the volatility of L, expressed by the Gaussian exponent σ^2.

Especially, if we consider the three particle system as a market model on its own, a Brownian motion as price process is unsatisfying, as we have already argued in the introductory chapter. A reasonable extension is the consideration of more general Lévy processes. In the last decade, Lévy processes have become a powerful tool to model more complex phenomena of price processes than it is possible with a Brownian motion; today, they already belong to the mainstream of financial mathematics (Schoutens (2003), Cont and Tankov (2004)). As an extension to the Brownian case, we consider α-stable Lévy processes with $\alpha \in (0, 2)$ as price processes in Section 3.4. There are two reasons for our choice: First, the distribution of the returns of α-stable Lévy processes is heavy-tailed, that is the probability to see a price change of a certain size decays only with a power law as function of the size. This feature has already been known for real price processes since the sixties (see Mandelbrot (1963))[b], but the same quantity is exponentially decreasing for a Brownian motion. Second, equally to a Brownian motion, α-stable processes are self-similar: If we scale time by c^α and space by c, the resulting process has the same distribution as the original one. Consequently, we can apply the same heuristic argument as for the Brownian motion above. It

---

[b]Although there are recent empirical studies suggesting that the power law behaviour does not hold for arbitrary large price changes, but only up to a certain size. The probability to see larger jumps seems to be exponentially decreasing (see Potters and Bouchaud (2003)).

turns out that the critical value for γ should be $\alpha - 1$. We prove in Section 3.4 that this conjecture holds. Additionally, we are able to show that we do not need the full power of self-similarity; it is basically sufficient if the tails have the right order. Theorem 3.10 states the main result for heavy-tailed Lévy processes.

For the price processes considered here, 1 is the highest value of γ for which the three particle model remains stable. Yet, we have only considered Markovian, that is memoryless, price processes, and stylized facts depending on memory effects (see Section 2.3) demonstrate that this class of processes does not capture the whole complexity of real price movements. Our results suggest that there is little hope that the market of the Opinion Game remains stable forever if γ is larger than 1. On the other hand, the appearance of an external drift that is strong enough to induce a gap size larger than the critical threshold seems to be a rare event for $\gamma \leq 1.5$. An analytical understanding of the (random) time to reach the critical threshold in dependence on γ is therefore an interesting and important issue for future research. Furthermore, it would be interesting to extend our results to non-Markovian price processes, since the critical γ-value is closely connected to the *typical* exit time of the price process from a tube. Then again, the exit time is heavily depending on the Hurst exponent H of that process. For α-stable Lévy processes, we have $H = 1/\alpha$. It is very well possible that the critical γ-value equals $1/H - 1$ for a much greater class of processes than we prove it for in this book. For a critical value being greater than 1.5, we then would need $H < 2/5$. Since stationary Markov processes always have a Hurst exponent of $1/2$ at least, considering price processes with a correlation structure is a reasonable approach, which is additionally covered by some empirical studies (Sang et al. (2001); Simonsen (2003)). They suggest that the price processes of real markets can have Hurst exponents smaller than $1/2$.

3.2 The construction of the particle system

The particle system consists of three particles whose positions are denoted by L, $X(L)$ and $Y(L)$. While L is an element of a certain subclass of Lévy processes,

3.2. THE CONSTRUCTION OF THE PARTICLE SYSTEM

which we specify later on, the dynamics of $X(L)$ and $Y(L)$ are basically described by ordinary differential equations that depend on L; the randomness in $X(L)$ and $Y(L)$ is consequently inherited from L. We define $X(L)$ and $Y(L)$ pathwise, and since the sample paths of Lévy processes are almost surely càdlàg, that is *continue à droite, limitée à gauche*, with finite left-hand limits, we start with a construction of $X(f)$ and $Y(f)$ that are attracted to a deterministic càdlàg function f with finite left-hand limits. The definitions of $X(L)$ and $Y(L)$ follow immediately from the deterministic case.

3.2.1 A deterministic approach...

We construct $X(f)$ by the following procedure: We consider a sequence of step functions (f^m) that converges to f. The construction of $X(f^m)$ turns out to be simple. We prove afterwards that $X(f^m)$ converges uniformly on every compact set to a limit process $X(f)$. The definition of $Y(f)$ is equivalent.

Let $f : [0, \infty) \to \mathbb{R}$ be a càdlàg function with $f(0) := 0$ that is bounded on every compact subset of $[0, \infty)$. The assumption of boundedness is equivalent to the assumption of finite left-hand limits. We define the *jump times* $(\tau_i^m)_{i \in \mathbb{N}_0}$ for all $m \in \mathbb{N}$ by $\tau_0^m := 0$ and

$$\tau_i^m := \min\left\{t > \tau_{i-1}^m : \left|f(\tau_{i-1}^m) - f(t)\right| \geq \frac{1}{m}\right\}, \quad i \in \mathbb{N}, \tag{3.2.1}$$

neglecting the m-index when no confusion is caused. Given the jump times, we construct the *step functions* $f^m : [0, \infty) \to \mathbb{R}$, $m \in \mathbb{N}$, by

$$f^m(t) := f(\tau_i) \text{ for } t \in [\tau_i, \tau_{i+1}). \tag{3.2.2}$$

(f^m) converges uniformly to f, because

$$\sup_{t \geq 0} |f(t) - f^m(t)| \leq \frac{1}{m}. \tag{3.2.3}$$

We basically want $X(f)$ to solve the ordinary differential equation

$$\frac{d}{dt}X_t = (K + f(t) - X_t)^{-\gamma}, \quad K, \gamma > 0, \tag{3.2.4}$$

as long as $X_t \leq f(t)$. If we substitute f by a constant $d \geq 0$, equation (3.2.4) is explicitly solvable. Namely, the solution of

$$\frac{d}{dt}g(t) = (K + d - g(t))^{-\gamma}, \quad g(0) = 0, \tag{3.2.5}$$

is

$$\bar{h}(t,d) := d + K - \left((d+K)^{\gamma+1} - (\gamma+1)t\right)^{\frac{1}{\gamma+1}}. \tag{3.2.6}$$

We call $\bar{h}(t,d)$ *well-defined* if

$$d \geq 0 \text{ and } t \leq \frac{(d+K)^{\gamma+1} - K}{\gamma+1}. \tag{3.2.7}$$

The bound on t ensures $\bar{h}(t,d) \leq d$. We are mainly interested in the distance from $\bar{h}$ to d at time t, thus we set

$$h(t,d) := \begin{cases} d - \bar{h}(t,d) & \text{if } \bar{h}(t,d) \text{ is well-defined} \\ 0 & \text{else} \end{cases}. \tag{3.2.8}$$

Now, we are able to define $X^m = (X(f^m)_t)_{t \geq 0}$ by

$$X_t^m := f^m(\tau_i) - h(t - \tau_i, f^m(\tau_i) - X_{\tau_i-}^m) \tag{3.2.9}$$

for $t \in [\tau_i, \tau_{i+1})$, $i \in \mathbb{N}_0$, and $X_{0-}^m := 0$. Expressed in words, we look at $X_{\tau_i-}^m$ first; if $f^m(\tau_i)$ is smaller than $X_{\tau_i-}^m$, we set $X_t^m := f^m(\tau_i)$, else we can apply the function h to calculate the movement of X^m towards f^m. If X^m hits f^m before time t, X^m remains on this level. Figure 3.5 illustrates the construction; the process $Y^m = (Y_t^m)_{t \geq 0}$ is defined by

$$Y_t^m := -X(-f^m)_t. \tag{3.2.10}$$

We prove next that (X^m) is a Cauchy sequence in the Banach space of bounded functions on compact sets, equipped with the supremum norm.

Lemma 3.1. *Let $S \subset [0, \infty)$ be a compact set, and let m and n be natural numbers*

3.2. THE CONSTRUCTION OF THE PARTICLE SYSTEM

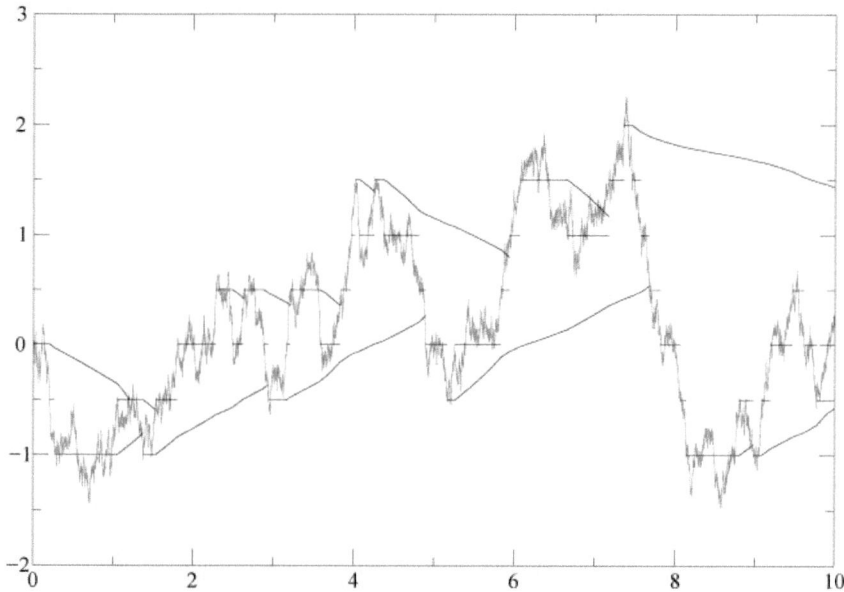

Figure 3.5: The three processes f^m (step function), X^m (lower function) and Y^m (upper function) for $m = 2$; here, f is the realisation of a Brownian motion and displayed beneath in a lighter grey.

with $m < n$. For sufficiently large m,

$$\sup_{t \in S} |X_t^m - X_t^n| \leq \frac{c}{m} \tag{3.2.11}$$

with c being a constant that does not depend on m.

In the following proof, we do not take advantage of the finer structure of X^m and X^n; our estimates are rough. However, more careful estimates would generally improve the constant c only, as there is no uniform convergence on $[0, \infty)$ for all f.

Proof. Since S is compact, we may assume that $S = [0, s]$ for some $s > 0$. For $t \geq 0$, we define the distances

$$a_t := m\left(X_t^m - X_t^n\right), \; b_t := f^m(t) - X_t^m, \; c_t := f^n(t) - f^m(t). \tag{3.2.12}$$

Observe that a/m is the variable we are interested in. We know by (3.2.3)

$$|c_t| \leq \frac{1}{m} + \frac{1}{n} < \frac{2}{m} \tag{3.2.13}$$

for all $t \geq 0$; furthermore,

$$\left|a_{\tau_1^m}\right| < 1, \tag{3.2.14}$$

because $X_t^m = 0$ and $X_t^n \in (-1/m, 1/m)$ for all $t \in [0, \tau_1^m)$.

We introduce a function $\bar{a} = (\bar{a}_t)_{t \geq 0}$ of upper bounds fulfilling

$$\bar{a}_t \geq \sup_{f \text{ càdlàg}} |a_t(f)|. \tag{3.2.15}$$

In particular, we focus on the sequence $(\bar{a}_{\tau_{k/m}})_{k \in \mathbb{N}}$, but we can extend $\bar{a}$ to the whole domain by linear interpolation, because $|a|$ grows linearly at most, and we construct $(\bar{a}_{\tau_{k/m}})_k$ in such a way that it is increasing. Our goal is the computation of $\bar{a}_s$.

We would like to know how fast $\bar{a}$ has to grow to dominate $\sup |a|$. To answer this question, observe that the slope of a is bounded from above by

$$\frac{d}{dt} a_t \leq m \left[(K + b_t)^{-\gamma} - \left(K + \frac{a_t}{m} + b_t + \frac{2}{m} \right)^{-\gamma} \right], \tag{3.2.16}$$

assuming that $a_t > 1$, which implies $b_t > 0$. The first term in the brackets describes the velocity of X^m, the second term the velocity of X^n; the inequality results from choosing the maximal value for c_t, $2/m$, as given by (3.2.13). Differentiation shows that the right-hand side is decreasing as a function of b_t, suggesting to choose b_t small for a fast growth of a. On the other hand, a cannot increase anymore once X^m has hit f^m. Thus, a grows the fastest if $b_t = 0$ and f^m increases linearly with slope $K^{-\gamma}$; then X^m also increases with its maximal velocity, $K^{-\gamma}$, and the speed of X^n is given by

$$\frac{d}{dt} X_t^n = \left(K + \frac{a_t}{m} + \frac{2}{m} \right)^{-\gamma}. \tag{3.2.17}$$

An equivalent bound can be found by considering $a_t < -1$; the case $a_t \in [-1, 1]$ is excluded for $t > 0$ by setting $\bar{a}_0 := 1$. In particular, we know that $|a_t|$ is

3.2. THE CONSTRUCTION OF THE PARTICLE SYSTEM

smaller than 1 for $t \leq 1/m$ (see (3.2.14) and the corresponding argumentation). An intuitive choice for $\bar{a}$ was consequently $\bar{a}_0 := 0$ and $\bar{a}_{1/m} := 1$. By setting $\bar{a}_0 := 1$, we just shift the indices. To determine $\bar{a}_{1/m}$, we solve

$$\frac{\bar{a}_0}{m} + \frac{K^{-\gamma}}{m} - \frac{\bar{a}_{1/m}}{m} \geq \left(K + \frac{\bar{a}_{1/m}}{m} + \frac{2}{m}\right)^{-\gamma} \frac{1}{m} \qquad (3.2.18)$$

On the left-hand side of (3.2.18), we calculate the increment of X^n in $[0, 1/m]$ by adding the starting distance to the increment of X^m, and subtracting the distance at $1/m$ afterwards. On the right-hand side, we have a lower bound for the increment of X^n, since we assume here that X^n has a constant velocity in $[0, 1/m]$ that depends on $\bar{a}_{1/m}$ only. Taylor expansion implies

$$\bar{a}_{1/m} \leq \bar{a}_0 + \frac{(\gamma + \delta) K^{-(\gamma+1)}}{m} (\bar{a}_0 + 2). \qquad (3.2.19)$$

The $\delta > 0$ captures all terms of order m^{-2} and smaller. Observe that we can use the same calculation to find $\bar{a}_{(k+1)/m}$ given $\bar{a}_{k/m}$. A proof by induction shows that the iterated application of (3.2.19) results in

$$\bar{a}_{k/m} \leq \bar{a}_0 + (\bar{a}_0 + 2) \sum_{j=1}^{k} \binom{k}{j} \left(\frac{(\gamma + \delta) K^{-(\gamma+1)}}{m}\right)^j. \qquad (3.2.20)$$

The series in (3.2.20) can be calculated explicitly by the binomial theorem; for $k = sm$, we get

$$\sup_{t \in [0,s]} |a_t| \leq \bar{a}_s \leq \bar{a}_0 + (\bar{a}_0 + 2) \left[e^{s(\gamma + \delta) K^{-(\gamma+1)}} - 1\right]. \qquad (3.2.21)$$

Since a is scaled by m, (3.2.21) concludes the proof. $\square$

Lemma 3.1 implies that (X^m) converges uniformly on S to a unique limit process. S is chosen arbitrarily, thus we can extend the construction to $[0, \infty)$. We denote the limit process by $X(f) = (X(f)_t)_{t \geq 0}$. Observe that $X(f)$ has the desired properties: $X_t \leq f(t)$ for all t, and $X(f)$ solves equation (3.2.4) as long as $X_t < f(t)$. The latter statement can formally be proven in the same fashion as Lemma 3.1.

We define $Y(f) = (Y(f)_t)_{t \geq 0}$ by

$$Y(f) := -X(-f). \tag{3.2.22}$$

A uniqueness result showing that the constructed processes, X and Y, are the only processes with the desired properties is forthcoming work.

3.2.2 ...applied to random paths

We consider a Lévy process $L := (L_t)_{t \geq 0}$ on a probability space $\{\Omega, \mathscr{F}, P\}$ that is equipped with the natural filtration of L denoted by $(\mathscr{F}_t)_{t \geq 0}$. As already mentioned, L can be chosen to be almost surely càdlàg (see Sato (2005), page 3); thus, we can find a set $\Omega_0 \subset \Omega$ with $P(\Omega_0) = 1$ and $L(\omega)$ being càdlàg for all $\omega \in \Omega_0$. For the sake of convenience, let us assume that $\Omega_0 = \Omega$. Then $X(L(\omega))$ and $Y(L(\omega))$ are well-defined for all $\omega \in \Omega$.

Observe that continuity of $L(\omega)$ implies continuity of $X(L(\omega))$ and $Y(L(\omega))$. This follows from the construction, as the corresponding step function $L(\omega)^m$ has only jumps of size $1/m$. Consequently, the jumps of $X(L(\omega)^m)$ and $Y(L(\omega)^m)$ can also not be larger than $1/m$ such that the jumps of $X(L(\omega))$ and $Y(L(\omega))$ have size

$$\lim_{m \to \infty} \frac{1}{m} = 0 \tag{3.2.23}$$

at most.

If $L(\omega)$ is not continuous, the jumps of $X(L(\omega))$ and $Y(L(\omega))$ have always the same sign each:

$$X(L(\omega))_t - X(L(\omega))_{t-} \leq 0 \quad \text{and} \quad Y(L(\omega))_t - Y(L(\omega))_{t-} \geq 0 \tag{3.2.24}$$

for all $t \geq 0$, since $X(L(\omega))$ and $Y(L(\omega))$ have only discontinuities if they are *pushed* by $L(\omega)$.

3.3 Stability of the Brownian particle system

We consider a Brownian motion as driving process. We adhere to the standard notation and denote the process by B instead of L. As B is also a Lévy process, $X(B)$ and $Y(B)$ are well-defined; we often omit the argument B in the remainder of this section. Observe that B is almost surely continuous such that X and Y inherit this property, as already discussed in the last paragraph of Section 3.2.2. Furthermore, B has a scaling parameter $\sigma > 0$; in particular, B_t is normally distributed with mean 0 and variance $t\sigma^2$. We call the parameter σ^2 *Gaussian coefficient*.

3.3.1 The main result for the Brownian motion

We are interested in the long-term behaviour of the particle system, especially in the distance between X and Y. Equivalently, we can ask for the qualitative behaviour of the last exit time from an r-ball with respect to the $||\cdot||_1$-norm,

$$\theta_r(X,Y) := \sup\{t \geq 0 : |Y_t - X_t| \leq r\}. \tag{3.3.1}$$

The main theorem shows a dichotomy for θ that depends on γ and σ.

Theorem 3.2. *Let B be a Brownian motion with Gaussian coefficient σ^2, and let $X(B)$ and $Y(B)$ be the attracted processes as defined in Section 3.2.2. Furthermore, we denote the normal distribution function with mean 0 and variance 1 by F. Then,*

(i)
$$(\forall r > 0)\ \theta_r(X,Y) < \infty\ a.s., \tag{3.3.2}$$

for
$$(\gamma > 1)\ or\ (\gamma = 1\ and\ \sigma > \sqrt{2}); \tag{3.3.3}$$

(ii)
$$(\forall r > 0)\ \theta_r(X,Y) = \infty\ a.s. \tag{3.3.4}$$

for

$$(\gamma < 1) \;\; or \;\; \left(\gamma = 1 \; and \; \sigma < \frac{1}{\sqrt{2}F^{-1}(7/8)}\right). \tag{3.3.5}$$

Observe that our results are completely independent of the constant K; also σ is only relevant in the critical case, $\gamma = 1$. Unfortunately, $F^{-1}(7/8) \approx 1.150$ such that we are not able to solve the critical case completely. It is possible that there exists a critical value σ^* such that the behaviour of $Y - X$ depends on K if $\gamma = 1$ and $\sigma = \sigma^*$; yet, we do not know how to prove or disprove this conjecture.

In the following sections, we write that $Y - X$ is *transient* if (3.3.2) holds; if (3.3.4) holds, we write $Y - X$ is *recurrent*.

3.3.2 Proof of Theorem 3.2.(i): the transient case

We begin with an overview of this section. We first prove transience of $Y^m - X^m$ for sufficiently large m, and we use two observations to decrease the complexity of the problem: First, we do not need to consider the absolute positions, X^m, B^m, Y^m, but it is sufficient to analyse the behaviour of $(B^m - X^m, Y^m - B^m)$. Second, the distance between X^m and Y^m can only increase at those times at which B^m has jumps. Thus, it is sufficient to consider $(B^m - X^m, Y^m - B^m)$ at the jump times of B^m only, resulting in a Markov chain Φ^m on $[0, \infty)^2$. We then spend some time on a detailed analysis of Φ^m: We explain what happens in the underlying process $(B^m - X^m, Y^m - B^m)$ in one step of Φ^m, take a closer look at the irreducibility properties of the Markov chain, and give a formal definition for transience of Φ^m. We argue afterwards that the transience of Φ^m implies the desired result, $\theta_r(X^m, Y^m) < \infty$ almost surely. Thanks to our preliminaries on Φ^m, we can then easily apply Theorem 3.4, which gives a criterion for the transience of Φ^m. In the end of the section, we argue that the transience of $Y^m - X^m$ implies the transience of $Y - X$.

Introducing Φ^m

Let us consider the two-dimensional process $(B^m - X^m, Y^m - B^m)$ and interpret it in the following as a particle moving in $[0, \infty)^2$. The randomness of B implies

3.3. STABILITY OF THE BROWNIAN PARTICLE SYSTEM

that the step functions (B^m) are also random, and that the jump times (τ_i) are stopping times. Observe that $Y^m - X^m$ is just the sum of both coordinates. Furthermore, because $Y^m - X^m$ can only increase at the times τ_i and decreases afterwards, we have

$$\inf_{t \in [\tau_i, \tau_{i+1})} (Y^m - X^m)_t = (Y^m - X^m)_{\tau_{i+1}-}. \tag{3.3.6}$$

For all $m \in \mathbb{N}$, we define a two-dimensional Markov chain $\Phi^m = (\Phi(B^m)_i)_{i \in \mathbb{N}}$ with state space $[0, \infty)^2$, equipped with the Borel-σ-algebra $\mathfrak{B}([0,\infty)^2)$, by

$$\Phi_i^m := (B^m - X^m, Y^m - B^m)_{\tau_i-} \tag{3.3.7}$$

with $\tau_0- := 0$. The j-step transition probabilities from $(x,y) \in [0,\infty)^2$ to $A \subset \mathfrak{B}([0,\infty)^2)$ are denoted by $P_{(x,y)}^j(A)$, but we neglect the index for $j = 1$. The generator L is given by

$$Lg(\bar{x}, \bar{y}) := \int_{[0,\infty)^2} P_{(\bar{x},\bar{y})}(d(x,y)) g(x,y) - g(\bar{x}, \bar{y}) \tag{3.3.8}$$

for suitable functions $g : [0,\infty)^2 \to [0,\infty)$.

Properties of Φ^m

In the following, it is of great importance to understand how the particle moves exactly while $\Phi_i^m = (x,y)$ jumps to Φ_{i+1}^m (Figure 3.6). At first, a jump of size $1/m$ happens at time τ_i. The position afterwards is either $(x + 1/m, (y - 1/m) \vee 0)$ or $((x - 1/m) \vee 0, y + 1/m)$ with probability $1/2$ each. Let us denote the new position by (x', y'). Before the next jump happens at time τ_{i+1}, the particle drifts into the origin's direction. If it reaches one of the axes, it remains there and drifts towards the other axis until it reaches $(0,0)$. Thus, the coordinates of Φ_{i+1}^m are given by $(h(\tau_{i+1} - \tau_i, x'), h(\tau_{i+1} - \tau_i, y'))$. Observe that Φ^m can only increase (in the $||\cdot||_1$-sense) on the axes.

We introduce

$$\bar{\tau}_i := \tau_{i+1} - \tau_i \stackrel{d}{=} \inf\left\{t > 0 : B_t = \frac{1}{m}\right\}. \tag{3.3.9}$$

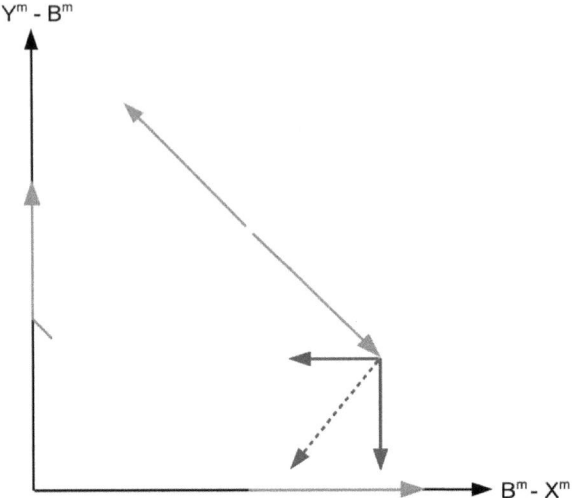

Figure 3.6: The particle's jumps (light grey arrows) are parallel to the level lines of the $\|\cdot\|_1$-norm. In the sense of this norm, the particle can only increase on the axes. The drift consists of two independent components (dark grey arrows), which are orthogonal to the axes. The resulting drift is illustrated by the dark grey, dashed blue arrow.

As already suggested by the equation above, all $\bar{\tau}_i$ are independent and identically distributed with support on $(0, \infty)$, and $\mathbb{E}\bar{\tau} = (m\sigma)^{-2}$. The distribution is not known explicitly, but it can be expressed as a series with alternating summands with decreasing absolute values (see Section C.2 in Berglund and Gentz (2006)). Calculating the first summand for the upper bound and, additionally, the second one for the lower bound results in

$$\frac{4}{\pi} \exp\left(-\frac{(\pi\sigma)^2 m}{8}\right) \left(1 - \frac{1}{3} e^{-(\pi\sigma)^2 m}\right)$$
$$\leq P\left(\bar{\tau} > 1/m\right) = P\left(\sup_{0 \leq s \leq 1/m} |B_s| < 1/m\right) \qquad (3.3.10)$$
$$\leq \frac{4}{\pi} \exp\left(-\frac{(\pi\sigma)^2 m}{8}\right).$$

For our purposes, it is sufficient to know that both bounds are of order $\exp(-m)$.

As we are operating on a continuous state space, the question for irreducibility is a question for reaching sets instead of single states. Formally, Φ^m is called

3.3. STABILITY OF THE BROWNIAN PARTICLE SYSTEM

φ-*irreducible* if there exists a measure φ on $\mathfrak{B}([0,\infty)^2)$ such that

$$\varphi(A) > 0 \Rightarrow P_{(x,y)}(\Phi^m \text{ ever reaches } A) > 0 \text{ for all } (x,y) \in [0,\infty)^2. \quad (3.3.11)$$

In our case,
$$P_{(x,y)}(\{\mathbf{0}\}) > 0 \text{ for all } (x,y) \in [0,\infty)^2, \quad (3.3.12)$$

because the support of $\bar{\tau}$'s density function is unbounded. Thus, Φ^m is δ_0-irreducible. The existence of an irreducibility measure ensures that there is also a *maximal irreducibility measure* Ψ (Proposition 4.2.2 in Meyn and Tweedie (1996)) on $\mathfrak{B}([0,\infty)^2)$ with the properties:

1. Ψ is a probability measure.

2. Φ^m is Ψ-irreducible.

3. Φ^m is φ'-irreducible iff $\Psi \succ \varphi'$.

4. $\Psi(A) = 0 \Rightarrow \Psi(\{(x,y) : P_{(x,y)}(\Phi^m \text{ ever enters } A)\}) = 0$.

5. Here, Ψ is equivalent to

$$\Psi'(A) = \sum_{j=0}^{\infty} P_0^j(A) 2^{-j}. \quad (3.3.13)$$

We denote the set of measurable, Ψ-irreducible sets by

$$\mathfrak{B}^+([0,\infty)^2) := \{A \in \mathfrak{B}([0,\infty)^2) : \Psi(A) > 0\}. \quad (3.3.14)$$

Because the density of $\bar{\tau}$ has support on $(0,\infty)$, it is easy to see that

$$\mu(A) := Leb(A) + \delta_0(A) \neq 0 \Rightarrow \Psi(A) \neq 0, \quad (3.3.15)$$

and therefore, $\Psi \succ \mu$ with *Leb* denoting the Lebesgue measure.

Transience of Φ^m and $Y^m - X^m$

We define transience of Φ^m as follows:

Definition 3.3. *For any $A \subset [0,\infty)^2$, let*

$$\eta_A := \sum_{i=0}^{\infty} \mathbb{1}_{\{\Phi_i^m \in A\}} \qquad (3.3.16)$$

be the number of visits of Φ^m in A. The set A is called uniformly transient if there exists $M < \infty$ such that $\mathbb{E}_{(x,y)}(\eta_A) \leq M$ for all $(x,y) \in A$. We call Φ^m transient if there is a countable cover of $[0,\infty)^2$ with uniformly transient sets.

Since Φ^m is a Markov chain on the *(possible) local minima* of $Y^m - X^m$ in the sense of (3.3.6), it is obvious that transience of Φ^m implies

$$\theta_r(X^m, Y^m) < \infty \text{ a.s.} \qquad (3.3.17)$$

On the other hand, (3.3.17) does not directly imply the desired result, (3.3.2), because we have convergence of (X^m, Y^m) to (X, Y) on compact sets only. We will return to this issue in the end of this section. We use the following theorem to show that Φ^m is transient in the upper sense. It can be found as a more general result in Meyn and Tweedie (1996), 8.0.2.(i), and it basically states that Φ^m is transient iff there is a bounded, subharmonic function with respect to the generator L.

Theorem 3.4. *The chain Φ^m is transient if and only if there exists a bounded, nonnegative function $g : [0,\infty)^2 \to [0,\infty)$ and a set $\mathcal{B} \in \mathfrak{B}^+([0,\infty)^2)$ such that, for all $(\bar{x}, \bar{y}) \in [0,\infty)^2 \setminus \mathcal{B}$,*

$$Lg(\bar{x}, \bar{y}) = \int_{[0,\infty]^2} P_{(\bar{x},\bar{y})}(d(x,y))g(x,y) - g(\bar{x},\bar{y}) \geq 0 \qquad (3.3.18)$$

and

$$D := \left\{ (x,y) \in [0,\infty)^2 \ : \ g(x,y) > \sup_{(\bar{x},\bar{y}) \in \mathcal{B}} g(\bar{x},\bar{y}) \right\} \in \mathfrak{B}^+([0,\infty)^2). \qquad (3.3.19)$$

We have to find a function g such that the particle moves away from the origin on average with respect to g. This must hold outside a compact set $\mathcal{B}$ containing

3.3. STABILITY OF THE BROWNIAN PARTICLE SYSTEM

the origin. To find a proper $\mathcal{B}$, we set

$$\mathcal{B}_z := \{(x,y) \in [0,\infty)^2 \,:\, \|(x+K,\, y+K)\|_{\gamma+1} = z\} \tag{3.3.20}$$

for all $z > 0$. For g, we choose

$$g(x,y) := 1 - \|(x+K,\, y+K)\|_{\gamma+1}^{-1}. \tag{3.3.21}$$

If we can find a $\bar{z}$ remaining finite as m tends to infinity such that condition (3.3.18) holds for all $(x,y) \in \mathcal{B}_z$, $z > \bar{z}$, we are done, since we can then set

$$\mathcal{B} := \bigcup_{z \leq \bar{z}} \mathcal{B}_z = \{(x,y) \in [0,\infty)^2 \,:\, \|(x+K,\, y+K)\|_{\gamma+1} \leq \bar{z}\}. \tag{3.3.22}$$

We can prove now that condition (3.3.18) holds under the assumptions of Theorem 3.2.(i). Recall what happens in one step of Φ^m in the underlying process as described on page 37. Equation (3.3.18) becomes

$$\begin{aligned}
&\frac{1}{2}\int_0^\infty P(\bar{\tau} \in dt)\, g\left(h\left(t, \bar{x}+\frac{1}{m}\right),\, h\left(t, \bar{y}-\frac{1}{m}\right)\right) \\
&+\frac{1}{2}\int_0^\infty P(\bar{\tau} \in dt)\, g\left(h\left(t, \bar{x}-\frac{1}{m}\right),\, h\left(t, \bar{y}+\frac{1}{m}\right)\right) \\
&\geq g(\bar{x}, \bar{y})
\end{aligned} \tag{3.3.23}$$

with $(\bar{x}, \bar{y}) \in \mathcal{B}_{\bar{z}}$. Because of the $1/m$-jump of B^m at time τ, the integral splits into two parts. Within both integrals, the only source of randomness is $\bar{\tau}$. If its value is given, we can calculate the next position of Φ^m and apply g to the new coordinates.

Using the definition of g and observing that the integral of the density $P(\bar{\tau} \in dt)$ is 1, we transform (3.3.23) to

$$\begin{aligned}
&\frac{1}{2}\int_0^\infty P(\bar{\tau} \in dt)\left\|\left(h\left(t, \bar{x}+\frac{1}{m}\right)+K,\, h\left(t, \bar{y}-\frac{1}{m}\right)+K\right)\right\|_{\gamma+1}^{-1} \\
&+\frac{1}{2}\int_0^\infty P(\bar{\tau} \in dt)\left\|\left(h\left(t, \bar{x}-\frac{1}{m}\right)+K,\, h\left(t, \bar{y}+\frac{1}{m}\right)+K\right)\right\|_{\gamma+1}^{-1} \\
&\leq \bar{z}^{-1}.
\end{aligned} \tag{3.3.24}$$

The drift time $\bar{\tau}$ is small in the sense of the inequalities (3.3.10); in particular, we can change the upper bounds of the integrals from ∞ to $1/m$ at the expense of order $\exp(-m)$. Furthermore, let us assume for the moment that $\bar{x}$ and $\bar{y}$ are large enough to avoid the case in which the axes are reached. Then, we can use Taylor approximations for $1/m$ and t to get

$$\frac{1}{2}\left[\left\|\left(h\left(t,\bar{x}+\frac{1}{m}\right)+K,\ h\left(t,\bar{y}-\frac{1}{m}\right)+K\right)\right\|_{\gamma+1}^{-1} \right. \tag{3.3.25}$$
$$\left. +\left\|\left(h\left(t,\bar{x}-\frac{1}{m}\right)+K,\ h\left(t,\bar{y}+\frac{1}{m}\right)+K\right)\right\|_{\gamma+1}^{-1}\right]$$
$$=\frac{1}{2}\left[\left(\left(\bar{x}+K+\frac{1}{m}\right)^{\gamma+1}+\left(\bar{y}+K-\frac{1}{m}\right)^{\gamma+1}-2(\gamma+1)t\right)^{-\frac{1}{\gamma+1}} \right. \tag{3.3.26}$$
$$\left. +\left(\left(\bar{x}+K-\frac{1}{m}\right)^{\gamma+1}+\left(\bar{y}+K+\frac{1}{m}\right)^{\gamma+1}-2(\gamma+1)t\right)^{-\frac{1}{\gamma+1}}\right]$$
$$=\bar{z}^{-1}+2\bar{z}^{-(\gamma+2)}t-\frac{\gamma}{2}\left((\bar{x}+K)^{\gamma-1}+(\bar{y}+K)^{\gamma-1}\right)\frac{\bar{z}^{-(\gamma+2)}}{m^2} \tag{3.3.27}$$
$$+(1+t)\mathcal{O}\left(\frac{\bar{z}^{-(2\gamma+3)}}{m^2}\right),$$

and, consequently, inequality (3.3.24) becomes

$$\int_0^{1/m} P(\bar{\tau}\in dt)\left[2\bar{z}^{-(\gamma+2)}t+(1+t)\mathcal{O}\left(\frac{\bar{z}^{-(2\gamma+3)}}{m^2}\right)\right]+\mathcal{O}(e^{-m})$$
$$\leq \frac{\gamma}{2}\left((\bar{x}+K)^{\gamma-1}+(\bar{y}+K)^{\gamma-1}\right)\frac{\bar{z}^{-(\gamma+2)}}{m^2}. \tag{3.3.28}$$

Because

$$\int_0^{1/m} P(\bar{\tau}\in dt)\, t \leq \mathbb{E}\bar{\tau} = (\sigma m)^{-2}, \tag{3.3.29}$$

we can rewrite (3.3.28) as

$$\frac{2}{\sigma^2}\frac{\bar{z}^{-(\gamma+2)}}{m^2}+\mathcal{O}\left(\frac{\bar{z}^{-(2\gamma+3)}}{m^2}\right)\leq \frac{\gamma}{2}\left((\bar{x}+K)^{\gamma-1}+(\bar{y}+K)^{\gamma-1}\right)\frac{\bar{z}^{-(\gamma+2)}}{m^2}, \tag{3.3.30}$$

which holds if

$$\gamma\left((\bar{x}+K)^{\gamma-1}+(\bar{y}+K)^{\gamma-1}\right) > \frac{4}{\sigma^2}. \tag{3.3.31}$$

Equation (3.3.31) is fulfilled for $\gamma > 1$ and $\bar{z}$ large enough, or for $\gamma = 1$ and $\sigma > \sqrt{2}$.

3.3. STABILITY OF THE BROWNIAN PARTICLE SYSTEM

Next, we show the special case that $\bar{x}$ or $\bar{y}$ is small enough for the particle to hit one of the axes. Because of symmetry, it is sufficient to consider one of the cases; we assume $\bar{x}$ to be small. Since the particle does a jump of size $1/m$ first, and drifts afterwards, $\bar{x}$ can only become 0 if

$$\bar{x} \in \left[0, \frac{1}{m} + \left(K^{\gamma+1} + (\gamma+1)\bar{\tau}\right)^{\frac{1}{\gamma+1}} - K\right]. \tag{3.3.32}$$

If we assume again that the drift time is $1/m$ at most, then

$$\bar{x} < \frac{1}{m} + K^{-\gamma}\bar{\tau}. \tag{3.3.33}$$

Condition (3.3.24) holds if

$$\frac{1}{2}\int_0^{1/m} P(\bar{\tau} \in dt) \left\| \left(h\left(t, \bar{x}+\frac{1}{m}\right)+K, \ h\left(t, \bar{y}-\frac{1}{m}\right)+K\right)\right\|_{\gamma+1}^{-1}$$
$$+\frac{1}{2}\int_0^{1/m} P(\bar{\tau} \in dt) \left\| \left(K, \ h\left(t, \bar{y}+\frac{1}{m}\right)+K\right)\right\|_{\gamma+1}^{-1} + \mathcal{O}(e^{-m}) \tag{3.3.34}$$
$$\leq \bar{z}^{-1}.$$

Applying Taylor approximation once again yields

$$\frac{1}{2}\left[\left\| \left(h\left(t, \bar{x}+\frac{1}{m}\right)+K, \ h\left(t, \bar{y}-\frac{1}{m}\right)+K\right)\right\|_{\gamma+1}^{-1} \right. \tag{3.3.35}$$
$$\left. + \left\| \left(K, \ h\left(t, \bar{y}+\frac{1}{m}\right)+K\right)\right\|_{\gamma+1}^{-1} \right]$$
$$= \bar{z}^{-1} + \frac{3}{2}\bar{z}^{-(\gamma+2)}t + \frac{1}{2}\left(K^\gamma \bar{x} + \frac{\gamma}{2}K^{\gamma-1}\bar{x}^2 - (\bar{x}+K)^\gamma \frac{1}{m}\right)\bar{z}^{-(\gamma+2)} \tag{3.3.36}$$
$$- \frac{\gamma}{2}\left(\frac{1}{2}(\bar{x}+K)^{\gamma-1} + (\bar{y}+K)^{\gamma-1}\right)\frac{\bar{z}^{-(\gamma+2)}}{m^2} + (1+t)\mathcal{O}\left(\frac{\bar{z}^{-(2\gamma+3)}}{m^2}\right).$$

Observe that

$$(3.3.27) = (3.3.36) + \frac{1}{2}\bar{z}^{-(\gamma+2)}t - \frac{\gamma}{4}(\bar{x}+K)^{\gamma-1}\frac{\bar{z}^{-(\gamma+2)}}{m^2}$$
$$- \frac{1}{2}\left(K^\gamma \bar{x} + \frac{\gamma}{2}K^{\gamma-1}\bar{x}^2 - (\bar{x}+K)^\gamma \frac{1}{m}\right)\bar{z}^{-(\gamma+2)}. \tag{3.3.37}$$

The additional terms on the right-hand side of the equation are increasing in $\bar{x}$ for $\gamma \geq 1$ and sufficiently large m as differentiation easily shows. We substitute $\bar{x}$

by its upper bound, $\frac{1}{m} + K^{-\gamma}\bar{\tau}$, given in line (3.3.33), to get

$$(3.3.27) \geq (3.3.36) - \frac{\gamma}{4}K^{-(\gamma+1)}z^{-(\gamma+2)}t^2 + \mathcal{O}\left(\frac{z^{-(\gamma+2)}}{m^3}\right). \qquad (3.3.38)$$

Equation (3.3.38) and the calculations before basically show that we benefit from the particle reaching 0 in one of its coordinates, as this coordinate cannot contribute to the particle's decrease (in $||\cdot||_{\gamma+1}$-sense) anymore. The closer to one of the axes the particle is, the greater the benefit is. Thus, we have a transition to the non-axes case, we have looked at before, when the particle reaches the axis at time $\bar{\tau}$ exactly. Then, both cases coincide (up to deviations of lower orders).

Transience of $Y^m - X^m$ implies transience of $Y - X$

It remains to argue that the transience of $Y^m - X^m$ as given in (3.3.17) implies the statement of the theorem, (3.3.2). For m large enough, we introduce the auxiliary processes $\hat{X}^m$ and $\hat{Y}^m$ that are constructed like X^m and Y^m but with a modified constant $\hat{K}$ instead of K:

$$\hat{K} := K - \frac{2}{m}. \qquad (3.3.39)$$

The crucial observation is that the *auxiliary* processes are sandwiched by the original processes:

$$X_t^n \leq \hat{X}_t^m + \frac{1}{m} \quad \text{and} \quad Y_t^n \geq \hat{Y}_t^m - \frac{1}{m} \qquad (3.3.40)$$

for all $n > m$ and all $t \geq 0$. This holds due to the fact that $|X_{\tau_1^n}^n - X_{\tau_1^m}^m| < 1/m$ and $|B^m - B^n| < 1/m$. Thus, the difference in speed cannot be larger than $2/m$. This argument extends to all later times τ_i inductively. We infer

$$\begin{aligned} Y_t - X_t &= \lim_{n\to\infty}(Y_t^n - X_t^n) & (3.3.41) \\ &\geq \hat{Y}_t^m - \hat{X}_t^m - \frac{2}{m} & (3.3.42) \end{aligned}$$

for all m. Since the proof also works for $\hat{K}$, we have obtained the desired result.

3.3.3 Proof of Theorem 3.2.(ii): the recurrent case

While the proof of the transient case is directly based on a generator argument, we use a coupling argument to prove Theorem 3.2.(ii). In particular, we consider $Y - X$ on a logarithmic scale at certain points in time. The resulting time discrete process is non-Markovian, since we do not take the position of B into consideration. Nevertheless, it is possible to construct a Markov chain on $\mathbb{Z}$ that dominates this process if coupled in the right way. It is sufficient to show then that the chain has a drift towards 0 for sufficiently large values as we show in Lemma 3.5. The existence of this drift under the assumptions of Theorem 3.2.(ii) follows then from the observation that the average time $Y - X$ needs to get from C^n to C^{n-1} for $C > 1$ and n large enough is shorter than the average time that a Brownian motion remains in a tube with diameter C^{n-1}.

Preliminaries

Let us consider $(D, E)(B) = ((D, E)(B)_t)_{t \geq 0}$ defined by

$$(D, E)(B)_t := (Y(B)_t - X(B)_t, B_t - X(B)_t). \tag{3.3.43}$$

(D, E) is a Markov process with state space $\{(x, y) \in [0, \infty)^2 : x \in [0, \infty), y \in [0, x]\}$; the Markov property is inherited from B. In contrast to (D, E), D is not Markovian, since the position of B is captured by E. Yet, D is irreducible, that is it has a positive probability to reach all sets $A \in \mathfrak{B}([0, \infty))$ with positive Lebesgue measure in finite time from any starting point in $[0, \infty)$. The irreducibility follows from the observation that D decreases with slope $2K^{-\gamma}$ at most, while the density of B_t has support on $[0, \infty)$ for all $t > 0$. Thus, there is a positive probability that B drives X arbitrarily far apart from Y. On the other hand, there is a positive probability for

$$\sup_{s \in [0,t]} |B_s| \leq \epsilon \tag{3.3.44}$$

for all positive, finite t and ϵ; if B remains in an ϵ-tube long enough, the distance between X and Y can decrease from any finite starting distance to 2ϵ.

We introduce the upper and lower exit time of D by

$$\nu_x^+(y)(B) := \min\{t \geq 0 : D(B)_t = y, D_0(B) = x\} \qquad (3.3.45)$$

for all $y \geq x \geq 0$, and

$$\nu_x^-(y)(B) := \min\{t \geq 0 : D(B)_t = y, D_0(B) = x\} \qquad (3.3.46)$$

for all $x \geq y > 0$. Next, we fix a constant $C > 1$, and define

$$p_n := \inf_{e \in [0, C^n]} P\left(\nu_{C^n}^-(C^{n-1}) < \nu_{C^n}^+(C^{n+1}) | E_0 = e\right) \qquad (3.3.47)$$

for all $n \in \mathbb{Z}$. The variable p_n denotes the minimal probability to leave $[C^{n-1}, C^{n+1}]$ at the lower boundary with respect to all possible values of E_0.

Reducing the problem to a question of drift

Lemma 3.5. *If there are constants $\delta > 0$ and $n^* \in \mathbb{Z}$ such that*

$$p_n > \frac{1}{2} + \delta \qquad (3.3.48)$$

for all $n \geq n^*$, then

$$(\forall r > 0) \; \theta_r(X, Y) = \infty \; \text{a.s.} \qquad (3.3.49)$$

For the proof of Lemma 3.5, we need the following Theorem.

Theorem 3.6 (Asmussen (1987), Proposition 1.5.3.(i))**.** *Consider an irreducible Markov chain with discrete state space $\mathcal{E}$ and transition probabilities (p_{ij}), and let $\mathcal{E}_0$ be a finite subset of $\mathcal{E}$. Then the chain is recurrent if there exists a function $h : \mathcal{E} \to \mathbb{R}$ such that $\{i : h(i) < J\}$ is finite for each J and*

$$\sum_{j \in \mathcal{E}} p_{ij} h(j) \leq h(i), \; i \notin \mathcal{E}_0. \qquad (3.3.50)$$

Proof of Lemma 3.5. We set $r^* := C^{n^*}$. Observe that it is sufficient to show

$$\theta_{r^*}(X, Y) = \infty \text{ almost surely,} \qquad (3.3.51)$$

3.3. STABILITY OF THE BROWNIAN PARTICLE SYSTEM

since $Y - X = D$ is irreducible. Furthermore, the time until $Y - X$ started in C^n hits C^{n-1} or C^{n+1} is almost surely finite for the same reasons that we have already mentioned to argue the irreducibility of D. We set $t_0 := \nu_0^+(r^*)$, and define recursively

$$t_k := \min\left\{t > t_{k-1} \,:\, D_t \in \{D_{t_{k-1}}/C, C D_{t_{k-1}}\}\right\} \qquad (3.3.52)$$

for $k \in \mathbb{N}$. These are the times when D hits a neighbouring level of the form C^n. We introduce the random process $\Xi = (\Xi_k)_{k \in \mathbb{N}_0}$ with state space $\mathbb{Z}$; it is defined by

$$\Xi_k := \frac{\ln D_{t_k}}{\ln C}. \qquad (3.3.53)$$

Ξ starts in n^* and can only move with step size 1 by the almost sure continuity of D. Let

$$k^* := \min\{k \in \mathbb{N} \,:\, \Xi_k \leq n^*\}, \qquad (3.3.54)$$

and observe

$$P(k^* \text{ is finite}) = 1 \;\Rightarrow\; (3.3.51). \qquad (3.3.55)$$

Since Ξ is not Markovian, we introduce a nearest neighbour random walk $\bar{\Xi} = (\bar{\Xi}_k)_{k \in \mathbb{N}_0}$ with state space $\mathbb{Z}$ and $\bar{\Xi}_0 := n^*$. The transition probabilities are

$$P\left(\bar{\Xi}_{k+1} = n - 1 \,\big|\, \bar{\Xi}_k = n\right) := \frac{1}{2} + \delta \;\;\text{ and }\;\; P\left(\bar{\Xi}_{k+1} = n + 1 \,\big|\, \bar{\Xi}_k = n\right) := \frac{1}{2} - \delta \qquad (3.3.56)$$

for all $k \in \mathbb{N}_0$ and $n \in \mathbb{Z}$. Here, the variable δ is the same as in line (3.3.48). By (3.3.47), (3.3.48) and (3.3.56), we can couple Ξ and $\bar{\Xi}$ such that

$$\Xi_k \leq \bar{\Xi}_k \qquad (3.3.57)$$

almost surely for all $k \in \{0, \ldots, k^*\}$. Thus

$$P(\bar{k}^* \text{ is finite}) = 1 \;\Rightarrow\; (3.3.55) \qquad (3.3.58)$$

for

$$\bar{k}^* := \min\{k \in \mathbb{N} \,:\, \bar{\Xi}_k \leq n^*\}. \qquad (3.3.59)$$

Last, we use Theorem 3.6 to show that the left-hand side of implication (3.3.58) is true if (3.3.48) holds. As $\bar{\Xi}$ starts in n^* and stops if it is again in n^* (or below) again, it is sufficient to consider $\bar{\Xi}$ on $[n^* - 1, \infty) \cap \mathbb{Z}$ only. If δ is equal to $1/2$, we are done. Thus we assume without loss of generality that $\delta \in (0, 1/2)$, which implies the irreducibility of $\bar{\Xi}$. Choosing $\mathcal{E}_0 := \{n^* - 1\}$ and $h(i) := i$, and applying condition (3.3.50) to $\bar{\Xi}$ results in

$$(n-1)\left(\frac{1}{2} + \delta\right) + (n+1)\left(\frac{1}{2} - \delta\right) = n - 2\delta < n \qquad (3.3.60)$$

for all $n \geq n^*$. $\square$

Proving that condition (3.3.48) of Lemma 3.5 holds

It remains to show that Lemma 3.5 applies under the assumptions of Theorem 3.2.(ii). Let us assume that $D_0 = C^n$ and that B, started in $X_0 + e$, remains in a tube with radius C^{n-2}. If X and Y have reached the boundaries of the tube, their distance is obviously C^{n-1}. We define the hitting times of X and Y with the tube's boundaries by

$$T_n^X(e) := \begin{cases} \min\{t : X_t - X_0 \geq e - C^{n-2}\} & \text{for } e \in (C^{n-2}, C^n] \\ 0 & \text{for } e \in [0, C^{n-2}] \end{cases}, \qquad (3.3.61)$$

$$T_n^Y(e) := \begin{cases} \min\{t : Y_0 - Y_t \geq C^n - e - C^{n-2}\} & \text{for } e \in [0, C^n - C^{n-2}) \\ 0 & \text{for } e \in [C^n - C^{n-2}, C^n] \end{cases}. \qquad (3.3.62)$$

We cannot explicitly calculate T^X or T^Y, but B is assumed to stay in the tube, and therefore B's distance to the boundaries of the tube can be C^{n-1} at most. We can use formula (3.2.6), to find upper bounds for T^X and T^Y. In particular, for

$$\bar{T}_n^X(e) := \begin{cases} \frac{1}{\gamma+1}\left[(e + C^{n-2} + K)^{\gamma+1} - (e - C^{n-2} + K)^{\gamma+1}\right] & \text{for } e \in (C^{n-2}, C^n] \\ 0 & \text{for } e \in [0, C^{n-2}] \end{cases}$$

$$(3.3.63)$$

and

$$\bar{T}_n^Y(e) := \begin{cases} \frac{1}{\gamma+1}\left[(C^n - e + C^{n-2} + K)^{\gamma+1} \\ \quad - (C^n - e - C^{n-2} + K)^{\gamma+1}\right] & \text{for } e \in [0, C^n - C^{n-2}] \\ 0 & \text{for } e \in [C^n - C^{n-2}, C^n] \end{cases} \quad (3.3.64)$$

we have

$$T_n^X(e) \leq \bar{T}_n^X(e) \quad \text{and} \quad T_n^Y(e) \leq \bar{T}_n^Y(e). \qquad (3.3.65)$$

We set

$$T_n(e) := \max\left\{T_n^X(e), T_n^Y(e)\right\} \leq \max\left\{\bar{T}_n^X(e), \bar{T}_n^Y(e)\right\}, \qquad (3.3.66)$$

which can be bounded by

$$\sup_{e \in [0, C^n]} T_n(e) \qquad (3.3.67)$$

$$\leq \bar{T}_n^X(C^n) = \bar{T}_n^Y(0) \qquad (3.3.68)$$

$$= \frac{1}{\gamma+1}\left[\left(\frac{C^2+1}{C^2}C^n + K\right)^{\gamma+1} - \left(\frac{C^2-1}{C^2}C^n + K\right)^{\gamma+1}\right] \qquad (3.3.69)$$

$$=: T_n^* \qquad (3.3.70)$$

as differentiation of $\bar{T}_n^X(e)$ easily shows. Now, we can use the Brownian scaling property of B (see Revuz and Yor (1991), Proposition 1.10.(iii)),

$$B_t \stackrel{d}{=} \frac{1}{c} B_{tc^2} \qquad (3.3.71)$$

for all $t \geq 0$ and $c > 0$, to bound p_n from above by

$$p_n \geq \inf_{e \in [0, C^n]} P\left(\sup_{s \in [0, T_n(e)]} |B_s| \leq C^{n-2}\right) \qquad (3.3.72)$$

$$\geq P\left(\sup_{s \in [0, T_n^*]} |B_s| \leq C^{n-2}\right) \qquad (3.3.73)$$

$$= P\left(\sup_{s \in [0,1]} |B_s| \leq \frac{C^n}{C^2 \sqrt{T_n^*}}\right). \qquad (3.3.74)$$

We can compute the fraction from line (3.3.74) and get

$$\frac{C^n}{C^2 \sqrt{T_n^*}} = C^{n\left(1-\frac{\gamma+1}{2}\right)} \frac{(\gamma+1)^{1/2}}{C^2 \left(\left(\frac{C^2+1}{C^2}\right)^{\gamma+1} - \left(\frac{C^2-1}{C^2}\right)^{\gamma+1}\right)^{1/2}} + \mathcal{O}\left(C^{-n\frac{\gamma+1}{2}}\right). \quad (3.3.75)$$

For $\gamma > 1$, this expression tends to infinity, and thus p_n tends to 1.

The critical case, $\gamma = 1$, reveals the relevance of the constant C; we could have also worked with a concrete value up to now. For $\gamma = 1$, expression (3.3.75) becomes

$$\frac{C^n}{C^2 \sqrt{T_n^*}} = \frac{1}{\sqrt{2}C} + \mathcal{O}\left(C^{-n}\right). \quad (3.3.76)$$

We write B^σ to stress that B_1^σ has variance σ^2; then,

$$p_n \geq P\left(\sup_{s \in [0,1]} |B_s^\sigma| < \frac{1}{\sqrt{2}C}\right) \quad (3.3.77)$$

$$= P\left(\sup_{s \in [0,1]} |B_s^1| < \frac{1}{\sqrt{2}C\sigma}\right) \quad (3.3.78)$$

$$\geq 4P\left(B_1^1 < \frac{1}{\sqrt{2}C\sigma}\right) - 3. \quad (3.3.79)$$

We have used

$$P\left(\sup_{s \in [0,t]} |B_s^1| \geq x\right) \leq 4P\left(B_t^1 \geq x\right) \quad (3.3.80)$$

for the last transformation (see Section 2.8.A in Karatzas and Shreve (1998)). Line (3.3.79) implies

$$P\left(B_1^1 < \frac{1}{\sqrt{2}C\sigma}\right) \geq \frac{7+2\delta}{8} \Rightarrow p_n \geq \frac{1}{2} + \delta. \quad (3.3.81)$$

B_1^1 is normally distributed with variance 1. With F being the normal distribution function, we can transform implication (3.3.81) to

$$\sigma < \frac{1}{C\sqrt{2}F^{-1}(7/8)} \Rightarrow p_n \geq \frac{1}{2} + \delta \text{ for some } \delta > 0. \quad (3.3.82)$$

As (3.3.82) has to hold for some $C > 1$ only, the proof is finished.

Let us remark that the analytical bound in (3.3.79) for the term (3.3.78) is quite good. Using this bound, we get

$$\sigma < \frac{1}{\sqrt{2}F^{-1}(7/8)} \approx 0.615. \qquad (3.3.83)$$

It is also possible to express term (3.3.78) as *inverse Laplace transform* $\mathcal{L}_s^{-1}$ with respect to s (see Borodin and Salminen (1996), formula II.1.3.0.2):

$$P\left(\sup_{s\in[0,1]} |B_s^1| < \frac{1}{\sqrt{2}C\sigma}\right) = \mathcal{L}_s^{-1}\left(\left(e^{\frac{s}{C\sigma}} + e^{-\frac{s}{C\sigma}}\right)^{-1}\right). \qquad (3.3.84)$$

The right-hand side of this term can be solved numerically. The resulting upper bound on σ is equal to (3.3.83) with an error smaller than 10^{-4}.

3.4 An analogy for heavy-tailed Lévy processes

We extend our results from the last section to a class of more general Lévy processes, which we specify next.

3.4.1 $\mathfrak{C}(\alpha)$ - A class of Lévy processes

Every Lévy process L can be characterised by its characteristic function

$$\mathbb{E}e^{izL_t} = \exp\left[t\left(i\mu z - \frac{z^2\sigma^2}{2} + \int_{\mathbb{R}}\left(e^{izx} - 1 - izx\mathbb{1}_{\{|x|<1\}}\right)\lambda(dx)\right)\right] \qquad (3.4.1)$$

with *drift term* $\mu \in \mathbb{R}$, *Gaussian coefficient* $\sigma^2 \geq 0$, and *Lévy measure* λ, which fulfils

$$\lambda(\{0\}) = 0 \quad \text{and} \quad \int_{\mathbb{R}}(1 \wedge x^2)\lambda(dx) < \infty. \qquad (3.4.2)$$

This characterisation is unique up to identity in law and is a consequence of the *Lévy-Khintchine formula* for infinitely divisible distributions (Theorem 8.1 and Corollary 11.6 in Sato (2005)). We call (μ, σ^2, λ) the *characteristic triplet* and

$$\Psi(z) := i\mu z - \frac{z^2\sigma^2}{2} + \int_{\mathbb{R}}\left(e^{izx} - 1 - izx\mathbb{1}_{\{|x|<1\}}\right)\lambda(dx), \ z \in \mathbb{R} \qquad (3.4.3)$$

the *characteristic exponent*. The famous *Lévy-Itô decomposition* states that every term in Ψ represents an own random process.

Theorem 3.7 (Lévy-Itô decomposition, Theorem 6.1 in Papapantoleon (2008)). *Let (μ, σ^2, λ) be a characteristic triplet. Then there is a probability space on which four mutually independent Levy processes $L^{(1)}$, $L^{(2)}$, $L^{(3)}$ and $L^{(4)}$ exist. $L^{(1)}$ is a constant drift, $L^{(2)}$ is a Brownian motion, $L^{(3)}$ is a compound Poisson process and $L^{(4)}$ is a square integrable pure jump martingale with an almost surely countable number of jumps on each finite time interval of magnitude[c] less than 1. Taking $L = L^{(1)} + L^{(2)} + L^{(3)} + L^{(4)}$, we have that there exists a probability space on which a Lévy process $L = (L_t)_{t \geq 0}$ with the given characteristic exponent is defined.*

It is possible to choose the $L^{(i)}$ in this way that μ corresponds to the slope of $L^{(1)}$, σ^2 is the Gaussian coefficient of $L^{(2)}$, and λ is the intensity measure of a *Poisson random measure* that determines the occurrence of jumps (Theorem 19.2 in Sato (2005)) for $L^{(3)}$ and $L^{(4)}$. In particular, the time to see a jump of magnitude dx is exponentially distributed with parameter $\lambda(dx)$.

We can now specify those Lévy processes that we consider in the remainder of this section; we begin with the definition of the particular Lévy measures:

Definition 3.8. *A measure λ is an element of the set of heavy-tailed Lévy measures $\mathfrak{C}_{meas}(\alpha)$, $\alpha \in (0, 2)$, iff*

1. *λ is a Lévy measure.*

2. *There exist two constants $c^+ > c^- > 0$ such that*

$$\frac{c^-}{x^\alpha} \leq \lambda([x, \infty)) \leq \frac{c^+}{x^\alpha} \text{ for } x \to \infty. \qquad (3.4.4)$$

and

$$\frac{c^-}{|x|^\alpha} \leq \lambda((-\infty, x]) \leq \frac{c^+}{|x|^\alpha} \text{ for } x \to -\infty. \qquad (3.4.5)$$

3. *The functions $\lambda((-\infty, -x])$ and $\lambda([x, \infty))$ are continuous for sufficiently large values of x.*

The assumption of continuity is not fundamental for our results, but it saves us from discussing special cases in the proof. In the next definition, we use $L(\mu, \sigma^2, \lambda)$ as a short notation for a Lévy process L with generating triplet (μ, σ^2, λ).

---
[c]By the *magnitude of a jump*, we always denote the absolute value of the jump size.

3.4. AN ANALOGY FOR HEAVY-TAILED LÉVY PROCESSES

Definition 3.9. *The class of heavy-tailed Lévy processes $\mathfrak{C}(\alpha)$ is given by*

$$\mathfrak{C}(\alpha) := \{L(\mu, 0, \lambda) \ : \ \lambda \in \mathfrak{C}_{meas}(\alpha), \ \mu \in \mathbb{R}\} \tag{3.4.6}$$

for $\alpha \in (0,1]$, and by

$$\mathfrak{C}(\alpha) := \left\{L(\mu, 0, \lambda) \ : \ \lambda \in \mathfrak{C}_{meas}(\alpha), \ \mu = -\int_{\mathbb{R}} x \mathbb{1}_{\{|x| \geq 1\}} \lambda(dx)\right\} \tag{3.4.7}$$

for $\alpha \in (1,2)$.

The set $\mathfrak{C}(\alpha)$ contains those Lévy processes that have no Brownian part but heavy tails in the Lévy measure. In the case $\alpha \in (1,2)$, the drift μ corrects the jumps with magnitudes larger than 1 in expectation such that the considered processes are martingales. We do not need such a condition for the other case, $\alpha \in (0,1]$, since there the linear drift is dominated by the jumps as we will see later. Observe that $\mathfrak{C}(\alpha)$ contains all *(strictly) α-stable Lévy processes* that have both positive and negative jumps. These processes have Lévy measures of the form

$$\lambda(dx) = \left(\frac{K^+}{x^{\alpha+1}} \mathbb{1}_{x>0} + \frac{K^-}{|x|^{\alpha+1}} \mathbb{1}_{x<0}\right) dx, \ K^+, K^- > 0, \tag{3.4.8}$$

for $\alpha \in (0,2)\setminus\{1\}$, and

$$\lambda(dx) = \frac{K}{x^2} dx, \ K > 0, \tag{3.4.9}$$

for $\alpha = 1$. The corresponding value of the drift is

$$\mu = \frac{K^- - K^+}{\alpha - 1} \tag{3.4.10}$$

for $\alpha \in (0,2)\setminus\{1\}$, and $\mu \in \mathbb{R}$ for $\alpha = 1$. One can check that the α-stable processes have the scaling property

$$L_t \stackrel{d}{=} \frac{1}{c} L_{tc^\alpha} \tag{3.4.11}$$

for all $t \geq 0$ and $c > 0$ (see Sections 13 and 14 in Sato (2005)). Observe that (3.4.11) is the counterpart to (3.3.71); thus a Brownian motion is a 2-stable Lévy process, which is the mathematical motivation to consider this class of Lévy processes as an extension of Theorem 3.2.

3.4.2 The main result for heavy-tailed Lévy processes

We are again interested in the long-term behaviour of $Y(L) - X(L)$, but this time for $L \in \mathfrak{C}(\alpha)$. Recall that $\theta_r(X, Y)$, defined in (3.3.1), denotes the last exit time of $Y - X$ from an r-ball.

Theorem 3.10. *Let L be an element of $\mathfrak{C}(\alpha)$, and let $X(L)$, $Y(L)$ be the attracted processes as defined in Section 3.2.2. Then,*

(i)
$$(\forall r > 0) \ \theta_r(X, Y) < \infty \ a.s., \tag{3.4.12}$$

for
$$(\gamma > \alpha - 1); \tag{3.4.13}$$

(ii)
$$(\exists \bar{r} > 0) \, (\forall r \geq \bar{r}) \ \theta_r(X, Y) = \infty \ a.s. \tag{3.4.14}$$

for
$$(\gamma < \alpha - 1). \tag{3.4.15}$$

Observe that $Y - X$ can never be recurrent for $\alpha \leq 1$, as γ is assumed to be positive. Comparing Theorem 3.10 to Theorem 3.2, we see that the theorems basically complement each other. However, Theorem 3.10 differs from the Brownian case in two points: first, we do not make any statement about the critical case, $\gamma = \alpha - 1$; second, the recurrence property in (ii) is weakened. Both changes are tributes to the higher degree of freedom in the class of considered processes. For a given characteristic triplet $(\mu, 0, \lambda)$, it is generally possible to strengthen the statements.

3.4.3 Preliminaries for both parts of the proof

We prove the result in the following two sections. The next section contains the proof of Theorem 3.10.(i); the section afterwards is devoted to the proof of 3.10.(ii). Both parts use the same idea as the recurrent case for the Brownian setting in Section 3.3.3: We consider again $Y - X$ on a logarithmic scale at certain points in time, which results in a non-Markovian process. We couple this process to a Markov chain with state space $\mathbb{Z}$ such that the Markov chain is almost surely smaller, respectively larger, than the non-Markovian process. Finally, it remains to show that the Markov chain has a drift to or away from 0 if the current state is large enough. It is mostly this last step, the proof of the existence of a drift, that makes the differences between the single cases (recurrent Brownian case, recurrent/transient Lévy case) and adds individual difficulties. We will see in the proof of transience for the Lévy case (Section 3.4.4) why it was simpler to use another idea to prove the Brownian complement. In all what follows, L is an element of $\mathfrak{C}(\alpha)$.

We begin with the formal generalisation of some definitions that we have already introduced for the Brownian case. $(D, E)(L) = ((D, E)(L)_t)_{t \geq 0}$ is given by

$$(D, E)(L)_t := (Y(L)_t - X(L)_t, L_t - Y(L)_t). \qquad (3.4.16)$$

$(D, E)(L)$ is also a Markov process with state space $\{(x, y) \in [0, \infty)^2 : x \in [0, \infty), y \in [0, x]\}$, since L is Markovian. The upper and lower exit times of D are defined by

$$\nu_x^+(y)(L) := \min\{t \geq 0 : D(L)_t \geq y, D_0(L) = x\} \qquad (3.4.17)$$

for all $y \geq x \geq 0$, and

$$\nu_x^-(y)(L) := \min\{t \geq 0 : D(L)_t = y, D_0(L) = x\} \qquad (3.4.18)$$

for all $x \geq y > 0$. Observe that it is still sufficient to consider "$D_t = y$" in (3.4.18), even though D is not continuous, since all discontinuities increase D because of observation (3.2.24); thus,

$$D_t - D_{t-} \geq 0 \qquad (3.4.19)$$

for all $t \geq 0$.

3.4.4 Proof of Theorem 3.10.(i): the transient case

More preliminaries

We fix a constant $C > 1$, and define

$$p_n^+ := \inf_{e \in [0,C^n]} P\left(\nu_{C^n}^+(C^{n+1}) < \nu_{C^n}^-(C^{n-1}) \mid E_0 = e\right), \qquad (3.4.20)$$

the minimal probability that D, started in C^n, leaves the interval (C^{n-1}, C^{n+1}) into the upper direction. Similar to the Brownian case, it is sufficient to prove that p_n^+ is larger than $1/2$.

Reducing the problem to a question of drift

Lemma 3.11. *If there are constants $\delta > 0$ and $n^* \in \mathbb{Z}$ such that*

$$p_n^+ > \frac{1}{2} + \delta \qquad (3.4.21)$$

for all $n \geq n^*$, then

$$(\forall r > 0) \ \theta_r(X, Y) < \infty \text{ a.s.,} \qquad (3.4.22)$$

For the proof of Lemma 3.4.20, we need the *complement* of Theorem 3.6.

Theorem 3.12 (Asmussen (1987), Proposition 1.5.4). *Consider an irreducible Markov chain with discrete state space $\mathcal{E}$ and transition probabilities (p_{ij}), and let $\mathcal{E}_0$ be a finite subset of $\mathcal{E}$. Suppose that*

$$\sum_{j \in \mathcal{E}} p_{ij} h(j) \leq h(i), \ i \notin \mathcal{E}_0, \qquad (3.4.23)$$

holds for some bounded h, satisfying $h(k) < h(l)$ for some $k \notin \mathcal{E}_0$ and all $l \in \mathcal{E}_0$. Then the chain is transient.

We deduced Theorem 3.4 from the more general Theorem 8.0.2.(i) in Meyn and Tweedie (1996). It should be possible to derive Theorem 3.6 in a similar way. However, the reference in Asmussen (1987) matches exactly our needs.

Proof of Lemma 3.11. We set $r^* := C^{n^*}$ and $t_0^+ := \nu_0^+(r^*)$. The time t_0^+ is almost

3.4. AN ANALOGY FOR HEAVY-TAILED LÉVY PROCESSES

surely finite, since $\mathfrak{C}(\alpha)$ only contains processes that have a positive probability for jumps of magnitude larger or equal than C^n for all n large enough; thus, $\nu_0^+(C^n)$ is almost surely positive for these n and, consequently, for all $n > 0$. The introduction of the largest power of C smaller than $x \in [0, \infty)$,

$$\lfloor x \rfloor := \max\{C^n : n \in \mathbb{Z},\ C^n \leq x\}, \qquad (3.4.24)$$

allows us to define

$$t_k^+ := \min\left\{t > t_{k-1}^+ : D_t \in \left(\lfloor D_{t_{k-1}^+}\rfloor/C\right] \cup \left[C\lfloor D_{t_{k-1}^+}\rfloor, \infty\right)\right\}, \quad k \in \mathbb{N}, \qquad (3.4.25)$$

the times when D leaves an interval of the form $[C^{n-1}, C^{n+1}]$. We introduce the random process $\Xi^+ = (\Xi_k^+)_{k \in \mathbb{N}_0}$ by

$$\Xi_k^+ := \frac{\ln D_{t_k^+}}{\ln C}. \qquad (3.4.26)$$

Ξ^+ has state space $\mathbb{R}$, and $\Xi_0^+ = \ln D_{t_0^+}/\ln C \geq n^*$. Ξ^+ is not Markovian, but we introduce the Markov chain $\bar{\Xi}^+ = (\bar{\Xi}_k^+)_{k \in \mathbb{N}_0}$ defined by $\bar{\Xi}_0^+ := n^*$ and the transition probabilities

$$P\left(\bar{\Xi}_{k+1}^+ = n-1 \mid \bar{\Xi}_k^+ = n\right) := \frac{1}{2} - \delta \quad\text{and}\quad P\left(\bar{\Xi}_{k+1}^+ = n+1 \mid \bar{\Xi}_k^+ = n\right) := \frac{1}{2} + \delta \qquad (3.4.27)$$

for all $k \in \mathbb{N}_0$ and $n \in \mathbb{Z}$, and with the same δ as in (3.4.21). By (3.4.20), (3.4.21) and (3.4.27), we can couple Ξ^+ and $\bar{\Xi}^+$ such that

$$\bar{\Xi}_k^+ \leq \Xi_k^+ \qquad (3.4.28)$$

almost surely for all $k \in \{0, \ldots, \bar{k}^*\}$ with

$$\bar{k}^* := \min\{k \in \mathbb{N} : \bar{\Xi}_k^+ \leq n^*\}. \qquad (3.4.29)$$

Let t^* be defined by

$$t^* := \min\{t \geq t_1^+ \mid D_t \leq C^{n^*}\}. \qquad (3.4.30)$$

Observe that $t^* = t_{k^*}^+$ for some $k^* \in \mathbb{N}$, since D is continuous when it decreases.

Similar to the Brownian recurrent case, the following implications hold

$$P(\bar{k}^* \text{ is infinite}) > 0 \Rightarrow P(k^* \text{ is infinite}) > 0 \Rightarrow P(t^* \text{ is infinite}) > 0 \Rightarrow (3.4.12).$$
(3.4.31)

We use Theorem 3.12 to prove

$$(3.4.21) \Rightarrow P(\bar{k}^* \text{ is infinite}) > 0. \tag{3.4.32}$$

We can restrict the state space of $\bar{\Xi}^+$ to $[n^* - 1, \infty) \cap \mathbb{Z}$, since we consider the chain only as long as it has not returned to a state smaller or equal to n^*. Furthermore, we may assume that $\bar{\Xi}^+$ is irreducible on $[n^*, \infty) \cap \mathbb{Z}$; else δ was equal to $1/2$, and this would imply transience anyway. For the irreducible Markov chain $\bar{\Xi}^+$, we set $h(i) := 1/i$ and $\mathcal{E}_0 := [n^* - 1, \max\{1/(2\delta), n^* - 1\}] \cap \mathbb{Z}$ in Theorem 3.12. Condition (3.4.23) becomes

$$\frac{\frac{1}{2} + \delta}{n+1} + \frac{\frac{1}{2} - \delta}{n-1} \leq \frac{1}{n}, \tag{3.4.33}$$

which transforms to

$$n \geq \frac{1}{2\delta}. \tag{3.4.34}$$

$\square$

Proving that condition (3.11) of Lemma 3.11 holds ...

It remains to show that Lemma 3.11 applies under the assumptions of Theorem 3.10.(i). We would like to find a lower bound for the probability that D started in C^n reaches $[C^{n+1}, \infty)$ before it hits C^{n-1}. The crucial observation that we use in the following calculations is

$$|L_t - L_{t-}| \geq C^{n+1} \Rightarrow D_t \geq C^{n+1}; \tag{3.4.35}$$

expressed in words, if the driving process L makes a jump of magnitude C^{n+1} at least, the distance between X and Y cannot be smaller than this magnitude in the moment after the jump has happened. We denote the time of the first jump

3.4. AN ANALOGY FOR HEAVY-TAILED LÉVY PROCESSES

with magnitude larger than, respectively equal to, $x > 0$ by

$$T_x := \min\{t \geq 0 \;:\; |L_t - L_{t-}| \geq x\}. \tag{3.4.36}$$

Recall that T_x is exponentially distributed with parameter $\lambda((-\infty, -x] \cup [x, \infty))$. We have to show that, with a high probability, D cannot reach C^{n-1} before time $T_{C^{n+1}}$. For $\alpha < 1$, this is a simple exercise, since D can decrease linearly at most, but $T_{C^{n+1}}$ increases sublinearly in n; we specify that statement in the next subsection. For $\alpha \geq 1$, we need some more work as we explain in the beginning of the corresponding subsection.

Before we begin our calculations, let us shortly remark that the idea above, to *compare* the time until L makes a large jump to the time that D needs to decrease under a given value, cannot be applied to the Brownian case, since a Brownian motion does not have jumps. This is the reason why we had to choose another way to prove the transient case with a Brownian motion as driving process in Section 3.3.3.

... for $\alpha \in (0,1)$ - a degeneracy

Let $\alpha \in (0,1)$. By the definition of $\mathfrak{C}(\alpha)$, there is a constant $c^- > 0$ such that

$$\int_{C^{n+1}}^{\infty} \lambda(dx) + \int_{-\infty}^{-C^{n+1}} \lambda(dx) \tag{3.4.37}$$

$$\geq \int_{C^{n+1}}^{\infty} \frac{c^-}{x^{\alpha+1}} dx + \int_{-\infty}^{-C^{n+1}} \frac{c^-}{|x|^{\alpha+1}} dx \tag{3.4.38}$$

$$= \frac{2c^-}{\alpha C^\alpha} C^{-n\alpha} \tag{3.4.39}$$

for n large enough. Furthermore, the velocities of X and Y are bounded from above by $K^{-\gamma}$. Thus, the minimal time that D needs to get from C^n to C^{n-1} is bounded from below by the time t^* that solves

$$C^n - 2K^{-\gamma} t^* = C^{n-1} \Leftrightarrow t^* = \frac{K^\gamma (1 - C^{-1})}{2} C^n. \tag{3.4.40}$$

We compute the probability to see a large jump before time t^*:

$$p_n^+ \geq P(T_{C^{n+1}} < t^*) \tag{3.4.41}$$

$$\geq 1 - \exp\left(-\frac{2c^-}{\alpha} C^{-(n+1)\alpha} t^*\right) \tag{3.4.42}$$

$$= 1 - \exp\left(-\frac{c^- K^\gamma \left(1 - C^{-1}\right)}{\alpha} C^{(1-\alpha)n}\right). \tag{3.4.43}$$

The expression in the last line tends to 1 as n tends to infinity. Thus, p_n^+ is larger than $1/2 + \delta$ for n large enough fulfilling the assumptions of Lemma 3.11; consequently, Theorem 3.10.(i) is proven for $\alpha \in (0, 1)$.

... for $\alpha \in [1, 2)$ - the sophisticated sibling

Let us explain the main difficulty of the case $\alpha \in [1, 2)$ in a non-rigorous way first: Let n be large, and assume $X_0 = 0$ and $Y_0 = C^n$ such that $D_0 = C^n$. We would like to show

$$\begin{aligned}p_n^+ &\geq \int_0^\infty P(T_{C^{n+1}} \in ds) \\ &\quad P\left(\left(\sup_{t \in [0,s]} X_t < \frac{1}{2} C^{n-1}\right) \wedge \left(\inf_{t \in [0,s]} Y_t > \frac{1}{2} C^{n-1}\right)\right) \end{aligned} \tag{3.4.44}$$

$$\geq \frac{1}{2} + \delta, \tag{3.4.45}$$

which basically means that X may not increase too much and Y may not decrease too much within the interval $[0, T_{C^{n+1}}]$. Now, let us assume that $L_t \in [b_n, C^n - b_n]$ for all $t \in [0, T_{C^{n+1}})$, that is L remains in a tube of diameter $C^n - 2b_n$ until it has a large jump. We would like to choose b_n in such a way that X and Y cannot reach the tube's boundaries before time $T_{C^{n+1}}$ (see Figure 3.7). Since the velocity of X is decreasing if the distance to L increases, we find a suitable b_n by solving

$$\bar{h}(T_{C^{n+1}}, b_n) := b_n + K - \left((b_n + K)^{\gamma+1} - (\gamma+1) T_{C^{n+1}}\right)^{\frac{1}{\gamma+1}} = b_n. \tag{3.4.46}$$

Here, we assume that X is attracted to the constant b_n, and we use formula (3.2.6), which describes the position of X if attracted to a constant level, to find the value of b_n that is reached at time $T_{C^{n+1}}$. Simple transformations show that

$$b_n := \left(K^{\gamma+1} + (\gamma+1) T_{C^{n+1}}\right)^{\frac{1}{\gamma+1}} - K \qquad (3.4.47)$$

is the solution of the equation above. Observe that $b_n \in \mathcal{O}(T_{C^{n+1}}^{\frac{1}{\gamma+1}})$. $T_{C^{n+1}}$ is of order $C^{n\alpha}$ in the sense that

$$1 - \exp\left(-\frac{2c^-}{\alpha C^\alpha}\eta\right) \leq P\left(T_{C^{n+1}} < \eta C^{n\alpha}\right) \leq 1 - \exp\left(-\frac{2c^+}{\alpha C^\alpha}\eta\right) \qquad (3.4.48)$$

for $\eta > 0$ and n large enough. Observe that both bounds are independent from n. The lower bound follows from the calculation of the mass that is contained in the tails of the Lévy measure in (3.4.37). The upper bound follows equivalently for a constant $c^+ > c^-$ by the properties of $\lambda \in \mathfrak{C}_{\text{meas}}(\alpha)$. In the same sense, let us think about b_n as being of order $C^{n\frac{\alpha}{\gamma+1}}$; then, $X_{T_{C^{n+1}}}$ is of order $C^{n\frac{\alpha}{\gamma+1}} < C^{n-1}/2$. Expressed in words, the movement of X, and equivalently of Y, until time $T_{C^{n+1}}$ is negligible with respect to the starting distance C^n, since $\alpha/(\gamma+1) < 1$ for $\gamma > \alpha - 1$, and we would be done if the made assumption, that L does not get close to X or Y before time $T_{C^{n+1}}$, held. However, the assumption does not hold in general. First, we have to consider all initial points of L such that L_0 can attain every value in $[0, C^n]$. Second, even if L starts within $[b_n, C^n - b_n]$, it typically leaves the tube before time $T_{C^{n+1}}$, since the definition of $T_{C^{n+1}}$ may restrict L to jumps with magnitude less than C^{n+1}, but there can still be jumps larger than $C^n - 2b_n$.

If L is close to X, there are basically two possible scenarios for D that could happen: First, D could increase, since it is possible that L pushes X down. On the other hand, X increases faster if it is close to L. Thus, it could also be possible that X increases faster than expected, *pulled* by a realisation of L that remains close by. Then, not only does D decrease, but it moves even faster than expected such that our argumentation from the last paragraph is possibly not valid anymore. Of course, the second scenario is unlikely, since L has too much fluctuations to remain close to X over a long time period without pushing it down. The rigorous proof that the second scenario does not happen requires nevertheless

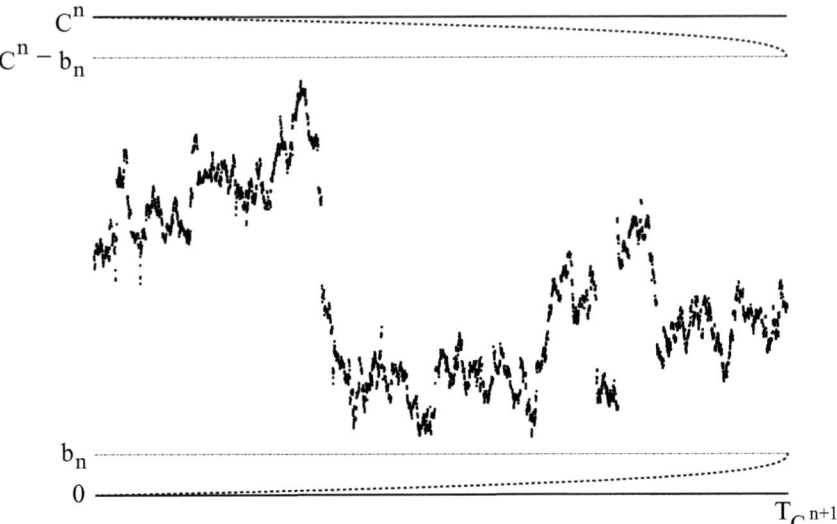

Figure 3.7: Sketch of the *simple* case. The black solid lines show the (almost static) dynamics of particles started in 0 or C^n and attracted to the constant levels b_n or $C^n - b_n$, respectively. If L (black with many jumps) remains in $[b_n, C^n - b_n]$ until time $T_{C^{n+1}}$, $X(L)_t$ is almost surely below the lower black, dashed line, and $Y(L)_t$ is almost surely above the upper black, dashed line for all $t \in [0, T_{C^{n+1}}]$.

some work.

We tackle this problem by cutting the time axis in *good* and *bad* subintervals, depending on if L is *close* to X (see Figure 3.8). We stitch together those parts of X that exist on the good time intervals and call the resulting process $\hat{X}$. By our argumentation from above, $\hat{X}$ cannot appreciably increase. Equally, we stitch together those parts of X that exist on the bad time intervals. We denote the resulting process by $\tilde{X}$. To show that $\tilde{X}$ also does not increase too much, we define a Lévy process $\tilde{X}^+$ that dominates $\tilde{X}$ from above. We can show then that $\tilde{X}^+$ can be decomposed into a martingale part and a strictly decreasing part. It suffices to show that the supremum of the martingale part does not increase too much with a high probability. Since $L \in \mathfrak{C}(\alpha)$ implies $-L \in \mathfrak{C}(\alpha)$, we can transfer our results to Y afterwards. In the remainder of this section, we do not mention again that we assume $\gamma > \alpha - 1$.

Recall that E denotes the distance between X and L. We set $v_{-1} := 0$ to define

$$u_k := \min \{ t \geq v_{k-1} \ : \ E_t \in [0, b_n] \} \qquad (3.4.49)$$

3.4. AN ANALOGY FOR HEAVY-TAILED LÉVY PROCESSES

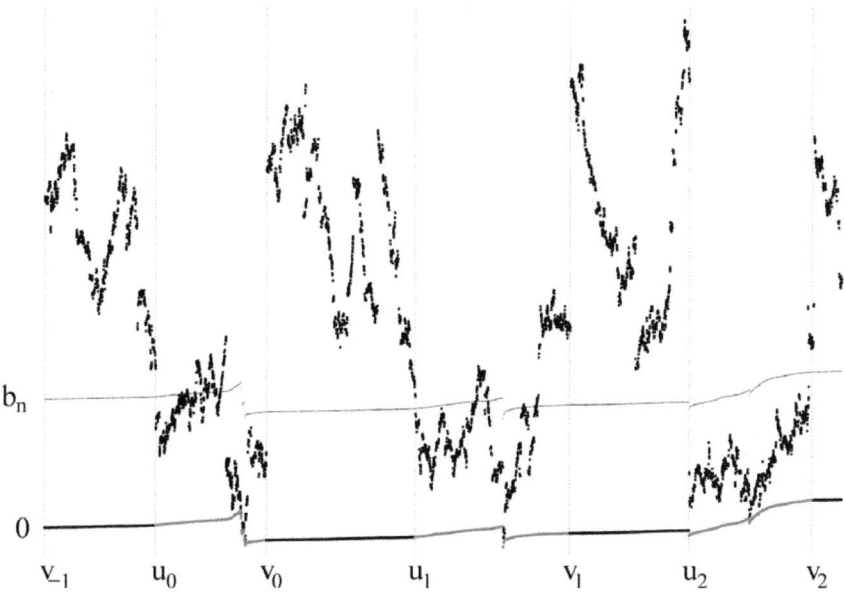

Figure 3.8: Sketch of a Lévy process L (black) and the attracted process $X(L)$ (black/light grey). The thin line has the same shape as X but is shifted by b_n. The black pieces of X lie in the good intervals, the grey pieces lie in the bad intervals. Observe that the distance between L and X can become larger than b_n in a bad interval.

and
$$v_k := \min\{t \geq u_k \,:\, (E_{t-} < b_n) \wedge (L_t - L_{t-} > b_n)\} \tag{3.4.50}$$

for $k \in \mathbb{N}_0$. $(u_k)_{k \in \mathbb{N}_0}$ denotes the times when the distance of L to X becomes smaller than b_n, $(v_k)_{k \in \mathbb{N}_0}$ denotes the times when the distance becomes larger than b_n again, but only if this happens by a jump of magnitude larger than b_n. These times are easier to control than the times when the distance just becomes larger than b_n.

No problems with the good intervals and $\hat{X}$. We define the *good* intervals $(I_k)_{k \in \mathbb{N}_0}$ by
$$I_k := [v_{k-1}, u_k) \tag{3.4.51}$$

and their length by
$$|I_k| := u_k - v_{k-1}. \tag{3.4.52}$$

Observe that I_0 is possibly not well-defined, since u_0 could be equal to $v_{-1} = 0$. In this case, let us assume that we do not consider I_0 and that we shift down all indices by 1. We stitch together the pieces of X in the intervals I_k to the process $\hat{X}$. We introduce a short notation for the sum of the interval lengths,

$$S(k) := \sum_{l=0}^{k} |I_l| \qquad (3.4.53)$$

for $k \in \mathbb{N}_0$, and define $\hat{X} = (\hat{X}_s)_{s \geq 0}$ by

$$\hat{X}_s := X_{s+v_{-1}} - X_{v_{-1}} \quad \text{for } s < |I_0|, \qquad (3.4.54)$$

and

$$\hat{X}_s := \hat{X}_{S(k-1)-} + X_{s-S(k-1)+v_{k-1}} - X_{v_{k-1}} \quad \text{for } S(k-1) \leq s < S(k). \qquad (3.4.55)$$

We use the time index s to emphasize that $\hat{X}$ has another time scale than X. $\hat{X}$ contains only pieces of X for which the distance to L is larger than b_n (see Figure 3.9), thus it is continuous and increasing. The amount of time that X spends in (I_k) until time $T_{C^{n+1}}$ is given by

$$\hat{s} := \left[\sum_{k=0}^{\infty} |I_k| \mathbb{1}_{\{u_k \leq T_{C^{n+1}}\}} \right] + (T_{C^{n+1}} - v_{k-1}) \mathbb{1}_{\{T_{C^{n+1}} \in I_k\}}. \qquad (3.4.56)$$

The increment of X within these intervals is then equal to $\hat{X}_{\hat{s}}$, and

$$\hat{X}_{\hat{s}} \leq \hat{X}_{T_{C^{n+1}}} \leq b_n \qquad (3.4.57)$$

by definition of $\hat{X}$ and b_n.

Defining the bad intervals and $\tilde{X}$. We consider the *bad* intervals:

$$J_k := [u_k, v_k) \qquad (3.4.58)$$

3.4. AN ANALOGY FOR HEAVY-TAILED LÉVY PROCESSES

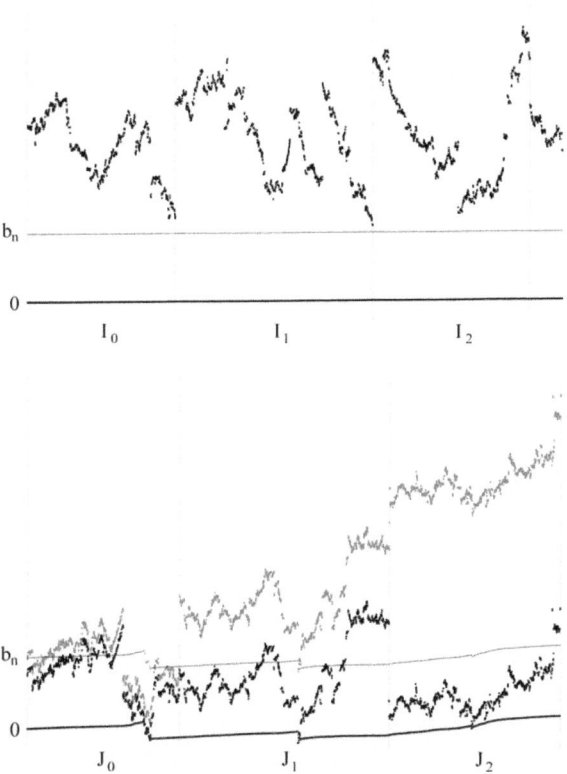

Figure 3.9: Sketch of L (black with jumps) and $\hat{X}(L)$ (black, bottom) in the upper graph, and L, $\tilde{X}(L)$ (black, bottom) and $\tilde{X}^+(L)$ (light grey) in the lower graph. Observe that $\tilde{X}$ is continuous at the boundaries of the intervals J_k; nevertheless, there can generally be discontinuities in the interior of the intervals. $\tilde{X}^+$ starts in b_n and is coupled to L. At the boundaries of the intervals J_k, it has jumps of size b_n. Since the distance of L to $\tilde{X}$ cannot be larger than b_n in the beginning of an interval J_k, and $\tilde{X}$ can never be larger than L (especially not in the end of the interval), $\tilde{X}^+$ dominates $\tilde{X}$.

for $k \in \mathbb{N}_0$. Again, let $|J|$ denote the length of the interval and $\bar{S}(k)$ the sum of $|J_0|$ to $|J_k|$. The process $\tilde{X} = (\tilde{X}_s)_{s \geq 0}$ is defined by

$$\tilde{X}_s := X_{s+u_0} - X_{u_0} \text{ for } s < |J_0|, \tag{3.4.59}$$

and

$$\tilde{X}_s := \tilde{X}_{\bar{S}(k-1)-} + X_{s-\bar{S}(k-1)+u_k} - X_{u_k} \quad \text{for } \bar{S}(k-1) \leq s < \bar{S}(k), \tag{3.4.60}$$

and is the union of those pieces of X that exist on the intervals (J_k) (see Figure 3.9). Observe that it is easily possible that X has discontinuities at the times u_k; this is the case if E_{u_k} is not only smaller than b_n but equal to 0 such that X is pushed down by L. On the other hand, $\tilde{X}$ is continuous at the u_k by construction; expressed in words, the additional pushes of L to X at u_k do not appear in $\tilde{X}$ although they decrease X and would therefore be helpful to show that X, and consequently $\tilde{X}$, cannot increase too much. Yet, we will see in the following that already $\tilde{X}$ as constructed above cannot appreciably increase; we just do not need to consider the additional pushes. The amount of time that X spends in (J_k) until time $T_{C^{n+1}}$ is given by

$$\tilde{s} := \left[\sum_{k=0}^{\infty} |J_k| \mathbb{1}_{\{v_k \leq T_{C^{n+1}}\}}\right] + (T_{C^{n+1}} - u_k) \mathbb{1}_{\{T_{C^{n+1}} \in J_k\}} \qquad (3.4.61)$$

Observe that $\tilde{s} \leq T_{C^{n+1}}$.

Defining $\tilde{X}^+$, a dominating process for $\tilde{X}$. By definition of $T_{C^{n+1}}$, L is conditioned on not having jumps of larger magnitude than C^{n+1} in $[0, T_{C^{n+1}})$; formally said, given L's characteristic triplet $(\mu, 0, \lambda)$, we can define another Lévy process $\bar{L} = (\bar{L}_t)_{t \geq 0}$ by its characteristic triplet $(\mu, 0, \bar{\lambda})$,

$$\bar{\lambda}(dx) := \mathbb{1}_{\{|x| < C^{n+1}\}} \lambda(dx), \qquad (3.4.62)$$

such that

$$\bar{L}_t \stackrel{d}{=} L_t \qquad (3.4.63)$$

for all $t \in [0, T_{C^{n+1}})$. Observe that $\tilde{X}$, different from $\hat{X}$, is not monotone. We would like to exclude that $\tilde{X}$ increases too much, thus we have to control it from above. We introduce a third process $\tilde{X}^+ = (\tilde{X}_s^+)_{s \geq 0}$ that dominates $\tilde{X}$ in the sense that $\tilde{X}^+(L)_s \geq \tilde{X}(L)_s$ almost surely for all $s \leq \tilde{s}$ (see Figure 3.9):

$$\tilde{X}_s^+ := b_n + \bar{L}_{s-u_0} - \bar{L}_{u_0} \text{ for } s < |J_0|, \qquad (3.4.64)$$

and

$$\tilde{X}_s^+ := b_n + \tilde{X}_{\bar{S}(k-1)-}^+ + \bar{L}_{s-\bar{S}(k-1)+u_k} - \bar{L}_{u_k} \quad \text{for } \bar{S}(k-1) \leq s < \bar{S}(k). \quad (3.4.65)$$

Observe $\tilde{X}$ starts in 0 and $\tilde{X}^+$ starts in b_n. Furthermore, $\tilde{X}^+$ is coupled to $\bar{L}$ by construction: The distance between the processes is piecewise constant. Since $\bar{L}_{u_0} - X_{u_0} \leq b_n$ by definition of u_0, and $X_t \leq \bar{L}_t$ for all $t \geq 0$, $\tilde{X}^+(\bar{L})_s$ dominates $\tilde{X}(\bar{L})_s$ for all $s < |J_0|$. We especially know

$$\tilde{X}^+(\bar{L})_{|J_0|-} \geq \tilde{X}(\bar{L})_{|J_0|-}. \quad (3.4.66)$$

Because

$$\tilde{X}(\bar{L})_{|J_0|} = \tilde{X}(\bar{L})_{|J_0|-} \quad (3.4.67)$$

by definition (3.4.61), it is reasonable to set

$$\tilde{X}^+(\bar{L})_{|J_0|} := \tilde{X}^+(\bar{L})_{|J_0|-} + b_n \geq \tilde{X}(\bar{L})_{|J_0|} + b_n. \quad (3.4.68)$$

Then, $\tilde{X}^+(\bar{L})_{|J_0|} - \tilde{X}(\bar{L})_{|J_0|} \geq b_n$ such that we can recursively apply the same argumentation as above. This procedure continues until time $\tilde{s}$.

$\tilde{X}^+ - b_n$ is a Lévy process. Observe that $\tilde{X}^+(\bar{L}) - b_n$ is a Lévy process. The (u_k) and the (v_k) are stopping times with respect to the natural filtration of $\bar{L}$, and $\bar{L}$ has the *strong Markov property* (Sato (2005), Corollary 40.11), thus $\bar{L}$ restarted in (u_k, L_{u_k}) is a Lévy process again. Furthermore, $\tilde{X}^+(\bar{L})$ has additional positive jumps of size b_n at the times (u_k), when $\bar{L}$ has jumps of size equal to or larger than b_n. We denote the rate of these jumps by

$$r_n := \int_{b_n}^{\infty} \bar{\lambda}(dx) = \int_{b_n}^{C^{n+1}} \lambda(dx). \quad (3.4.69)$$

This enables us to give the explicit characteristic exponent of $\tilde{X}^+ - b_n$:

$$\Psi_+(z) := i\mu z + \int_0^{b_n} \left(e^{izx} - 1 - izx \mathbb{1}_{\{x<1\}}\right) \lambda(dx) \quad (3.4.70)$$
$$+ \int_{-C^{n+1}}^0 \left(e^{izx} - 1 - izx \mathbb{1}_{\{|x|<1\}}\right) \lambda(dx) + \left(e^{ib_n z} - 1\right) r_n.$$

The last term incorporates the extra jumps of size b_n. We have chosen to write this term explicitly to emphasize it. Nevertheless, we could also include it in the integral by using the *Dirac measure* to obtain the Lévy-Khintchine formula.

$\tilde{X}^+$ is nicely decomposable. We claim that we can write $\tilde{X}^+$ as the decomposition

$$\tilde{X}_s^+ = M_s + A_s + b_n, \tag{3.4.71}$$

where $M = (M_t)_{t\geq 0}$ is a martingale and $A = (A_t)_{t\geq 0}$ is an almost surely decreasing process. Furthermore, M and A are Lévy processes. Our claim is proven if we can find a constant $\bar{b}_n < C^{n+1}$ such that

$$\frac{d}{dz}\Psi_M(z)\bigg|_{z=0} = 0 \tag{3.4.72}$$

for the characteristic function

$$\Psi_M(z) := i\mu z + \int_0^{b_n} \left(e^{izx} - 1 - izx\mathbb{1}_{\{x<1\}}\right)\lambda(dx) \tag{3.4.73}$$
$$+ \int_{-\bar{b}_n}^0 \left(e^{izx} - 1 - izx\mathbb{1}_{\{|x|<1\}}\right)\lambda(dx) + \left(e^{ib_n z} - 1\right)r_n$$

of M. Condition (3.4.72) guarantees the martingale property, and $\bar{b}_n < C^{n+1}$ implies that $A = \tilde{X}^+ - b_n - M$ consists of negative jumps only.

Let us first consider $\alpha \in (1,2)$. Recall that the definition of $\mathfrak{C}(\alpha)$ then determines the value of μ exactly. Taking also the definition of r_n in (3.4.69) into account transforms condition (3.4.72) into

$$\int_{\bar{b}_n}^\infty x\lambda(-dx) = \int_{b_n}^\infty x\lambda(dx) - b_n \int_{b_n}^{C^{n+1}} \lambda(dx) \tag{3.4.74}$$
$$= \int_{b_n}^{C^{n+1}} (x - b_n)\lambda(dx) + \int_{C^{n+1}}^\infty x\lambda(dx). \tag{3.4.75}$$

The right-hand side of the equation is positive and of order $b_n^{1-\alpha}$, because $\lambda \in \mathfrak{C}(\alpha)$. Consequently, the left-hand side is of order $\bar{b}_n^{1-\alpha}$ such that $\bar{b}_n$ must be of the same order than b_n, and therefore smaller than C^{n+1}.

3.4. AN ANALOGY FOR HEAVY-TAILED LÉVY PROCESSES

For $\alpha = 1$, we can transform condition (3.4.72) into

$$\int_1^{\bar{b}_n} x\lambda(-dx) = \mu + \int_1^{b_n} x\lambda(dx) + b_n \int_{b_n}^{C^{n+1}} \lambda(dx). \qquad (3.4.76)$$

The left-hand side is of order $\ln \bar{b}_n$, the right-hand side is of order $\ln b_n$. It follows again that b_n is smaller than C^{n+1}.

Reaping the fruits so far. Before we reap the fruits of our calculations, we cite *Doob's L^p-inequality*, which we need in the following.

Theorem 3.13 (Doob's L^p-inequality, Theorem II.1.7 in Revuz and Yor (1991)). *If M is a right-continuous martingale indexed by an interval $I \subset \mathbb{R}$, then if $M^* = \sup_{s \in I} |M_s|$, for $p \geq 1$,*

$$c^p P\left(M^* \geq c\right) \leq \sup_{s \in I} \mathbb{E}\left(|M_s|^p\right). \qquad (3.4.77)$$

Collecting all results from the previous pages yields a sequence of inequalities:

$$P\left(\sup_{t \in [0, T_{C^{n+1}}]} X_t \geq \frac{1}{2} C^{n-1}\right) \qquad (3.4.78)$$

$$= P\left(\hat{X}_{\tilde{s}} + \sup_{s \in [0, \tilde{s}]} \tilde{X}_s \geq \frac{1}{2} C^{n-1}\right) \qquad (3.4.79)$$

$$\overset{(3.4.57)}{\leq} P\left(\sup_{s \in [0, \tilde{s}]} \tilde{X}_s \geq \frac{1}{2} C^{n-1} - b_n\right) \qquad (3.4.80)$$

$$\leq P\left(\sup_{s \in [0, \tilde{s}]} \tilde{X}_s^+ \geq \frac{1}{2} C^{n-1} - b_n\right) \qquad (3.4.81)$$

$$\overset{(3.4.71)}{\leq} P\left(\sup_{s \in [0, \tilde{s}]} (M_s + A_s) \geq \frac{1}{2} \left(C^{n-1} - 4b_n\right)\right) \qquad (3.4.82)$$

$$\leq P\left(\sup_{s \in [0, T_{C^{n+1}}]} |M_s| \geq \frac{1}{2} \left(C^{n-1} - 4b_n\right)\right) \qquad (3.4.83)$$

$$\overset{\text{Thm. 3.13}}{\leq} \frac{\sup_{s \in [0, T_{C^{n+1}}]} \mathbb{E}\left(M_s^2\right)}{\left(\frac{1}{2}\left(C^{n-1} - 4b_n\right)\right)^2}. \qquad (3.4.84)$$

Line (3.4.81) follows, because $\tilde{X}^+$ dominates $\tilde{X}$ from above; line (3.4.83) follows, since A is almost surely decreasing, and $\tilde{s}$ is smaller than $T_{C^{n+1}}$.

The martingale part of $\tilde{X}^+$ does not fluctuate too much. We compute $\mathbb{E}(M_s^2)$. Since λ is an element of $\mathfrak{C}(\alpha)$, there exists an $x^*(\lambda) > 0$ such that

$$\frac{c^-}{x^\alpha} \leq \lambda([x,\infty)) \leq \frac{c^+}{x^\alpha} \quad \text{and} \quad \frac{c^-}{x^\alpha} \leq \lambda((-\infty,-x]) \leq \frac{c^+}{x^\alpha} \tag{3.4.85}$$

for all $x \geq x^*$. Furthermore, M is a Lévy process, thus

$$\mathbb{E}\left(M_s^2\right) \tag{3.4.86}$$

$$= -s \left.\frac{d^2}{dz^2}\Psi_M(z)\right|_{z=0} \tag{3.4.87}$$

$$\leq s\left[\int_0^{b_n} c^+ x^2 \lambda(dx) + \int_{-\bar{b}_n}^0 c^+ x^2 \lambda(dx) - b_n^2 \int_{b_n}^{C^{n+1}} \lambda(dx)\right] \tag{3.4.88}$$

$$\leq s\left[\mu^* + c^+ \int_{x^*}^{b_n} x^{1-\alpha} dx + c^+ \int_{x^*}^{\bar{b}_n} x^{1-\alpha} dx - c^+ b_n^2 \int_{b_n}^{C^{n+1}} x^{-(1+\alpha)} dx\right] \tag{3.4.89}$$

$$\in \Theta(s b_n^{2-\alpha}). \tag{3.4.90}$$

In line (3.4.89), we collect all integrals from 0 to $\pm x^*$ in μ^*, which depends on λ but not on n.

Reaping the remaining fruits. Recall that b_n is of order $T_{C^{n+1}}^{\frac{1}{\gamma+1}}$ (see (3.4.47)). Combining this observation with (3.4.90) and applying them to (3.4.84) yields

$$P\left(\sup_{t\in[0,T_{C^{n+1}}]} X_t \geq \frac{1}{2}C^{n-1}\right) \in \mathcal{O}\left(\frac{T_{C^{n+1}}^{1+\frac{2-\alpha}{\gamma+1}}}{C^{2n}}\right). \tag{3.4.91}$$

We finish the proof of transience by the following calculation:

$$p_n^+ \geq \int_0^\infty P\left(T_{C^{n+1}} \in ds\right)$$

$$P\left(\left(\sup_{t\in[0,s]} X_t < \frac{1}{2}C^{n-1}\right) \wedge \left(\inf_{t\in[0,s]} Y_t > \frac{1}{2}C^{n-1}\right)\right) \tag{3.4.92}$$

$$\geq \left(1 - \exp\left(-\frac{2c^-}{\alpha C^\alpha}\eta\right)\right)$$

$$P\left(\left(\sup_{t\in[0,\eta C^{n\alpha}]} X_t < \frac{1}{2}C^{n-1}\right) \wedge \left(\inf_{t\in[0,\eta C^{n\alpha}]} Y_t > \frac{1}{2}C^{n-1}\right)\right) \tag{3.4.93}$$

$$\geq \left(1 - \exp\left(-\frac{2c^-}{\alpha C^\alpha}\eta\right)\right)\left(1 - 2P\left(\sup_{t\in[0,\eta C^{n\alpha}]} X_t \geq \frac{1}{2}C^{n-1}\right)\right). \tag{3.4.94}$$

In line (3.4.94), we use that $P(A \cap B) = 1 - P(A^c \cup B^c) \geq 1 - P(A^c) - P(B^c)$ for two events A and B. Furthermore, the particle system's symmetry implies that we have the same upper bound for the probability of $\inf Y$ being smaller than $C^n - C^{n-1}/2$ as we have it for $\sup X$ being larger than $C^{n-1}/2$. Observe that $\eta > 0$ is a free parameter. By (3.4.91), we have

$$P\left(\sup_{t \in [0, \eta C^{n\alpha}]} X_t \geq \frac{1}{2} C^{n-1}\right) \in \mathcal{O}\left(C^{n\alpha + n\frac{\alpha(2-\alpha)}{\gamma+1} - 2n}\right). \quad (3.4.95)$$

A simple computation shows

$$n\alpha + n\frac{\alpha(2-\alpha)}{\gamma+1} - 2n < 0 \Leftrightarrow \gamma > \frac{(\alpha-1)(\alpha-2)}{\alpha-2}. \quad (3.4.96)$$

Consequently, Lemma 3.11 holds for $\gamma > \alpha - 1$.

3.4.5 Proof of Theorem 3.10.(ii): the recurrent case

We have already proven transience for all $\gamma > 0$ and $\alpha = 1$. Thus, we exclude $\alpha = 1$ and consider $\alpha \in (1, 2)$ only. The proof of recurrence works as in the Brownian case, but an additional difficulty is introduced, as the Markov chain that we introduce can have steps of arbitrary positive size. We use conditional probabilities to solve this problem.

More preliminaries

D is irreducible on $[\bar{r}, \infty)$ for $\bar{r} > 0$ large enough. In particular, D has a positive probability to reach every set $B \in \mathcal{B}([\bar{r}, \infty))$ with positive Lebesgue measure, since D is at least as large as the jumps of the underlying process L, and, by definition of $\mathfrak{C}(\alpha)$, the magnitude of these jumps is unbounded. On the other hand, L is a martingale, and thus there is a positive probability to remain in a tube of radius $\bar{r}/2$ for any finite time.

For a constant $C > 1$, we define $(p_{n,m}^-)$, $n \in \mathbb{Z}$, $m \in \{-1\} \cup \mathbb{N}$, by

$$p_{n,-1}^-(e) := P\left(\nu_{C^n}^-(C^{n-1}) < \nu_{C^n}^+(C^{n+1}) \mid E = e\right) \quad (3.4.97)$$

and, for $m \in \mathbb{N}$, by

$$p_{n,m}^-(e) := P\left(\nu_{C^n}^+(C^{n+1}) = \nu_{C^n}^+(C^{n+m}),\ \nu_{C^n}^+(C^{n+m}) < \nu_{C^n}^+(C^{n+m+1}),\right.$$
$$\left.\nu_{C^n}^+(C^{n+1}) < \nu_{C^n}^-(C^{n-1})\ \middle|\ E = e\right) \qquad (3.4.98)$$

for all $e \in [0, C^n]$. $p_{n,m}^-(e)$ is the probability that D started in C^n jumps to the interval $[C^{n+m}, C^{n+m+1})$ before it reaches any other interval of this form. The case "$m = -1$" is special, since D is continuous when it decreases. Observe that

$$p_{n,-1}^-(e) + \sum_{m=1}^{\infty} p_{n,m}^-(e) = 1 \qquad (3.4.99)$$

for all $n \in \mathbb{Z}$ and $e \in [0, C^n]$.

We define the probability to jump into $[C^{n+m}, C^{n+m+1})$ conditioned on not jumping in an interval below:

$$q_{n,-1}^-(e) := p_{n,-1}^-(e), \qquad (3.4.100)$$

$$q_{n,m}^-(e) := P\left(\nu_{C^n}^+(C^{n+m}) < \nu_{C^n}^+(C^{n+m+1})\ \middle|\right. \qquad (3.4.101)$$
$$\left. E = e,\ \nu_{C^n}^+(C^{n+m}) = \nu_{C^n}^+(C^{n+1}) < \nu_{C^n}^-(C^{n-1})\right)$$

for $n \in \mathbb{Z}$, $m \in \mathbb{N}$ and $e \in [0, C^n]$. Since

$$q_{n,m}^-(e) = \frac{p_{n,m}^-(e)}{\sum_{k=m}^{\infty} p_{n,k}^-(e)}, \qquad (3.4.102)$$

a proof by induction shows

$$p_{n,m}^-(e) = \left(1 - p_{n,-1}^-(e)\right) \left[\prod_{k=1}^{m-1}(1 - q_{n,k}^-(e))\right] q_{n,m}^-(e) \qquad (3.4.103)$$

for all $m \in \mathbb{N}$ and the term in the squared brackets being equal to 0 for $m = 1$. Equation (3.4.103) evidences that we can split the sampling of the next position of D in terms of $[C^{n+m}, C^{n+m+1})$-intervals into a sequence of Bernoulli trials. The first trial decides if D decreases below C^{n-1}; if it fails, the next trial decides if D increases to $[C^{n+1}, C^{n+2})$, and so on, until a trial is successful.

3.4. AN ANALOGY FOR HEAVY-TAILED LÉVY PROCESSES

For the remainder of this chapter, it is useful to introduce the short notation

$$p^-_{n,-1} := \inf_{e \in [0, C^n]} p^-_{n,-1}(e) \tag{3.4.104}$$

for all $n \in \mathbb{Z}$.

Reducing the problem to a question of drift

Lemma 3.14. *If there are constants $q \in (0, 1]$, $\delta \leq q/(1 + 2q)$ and $n^* \in \mathbb{Z}$ such that*

$$(\forall m \geq 2) \quad q \leq \inf_{e \in [0, C^n]} q^-_{n,m}(e) \tag{3.4.105}$$

and

$$p^-_{n,-1} \geq (1 - \delta), \tag{3.4.106}$$

for all $n \geq n^*$, then

$$(\exists \bar{r} > 0) \, (\forall r \geq \bar{r}) \quad \theta_r(X, Y) = \infty \text{ a.s.} \tag{3.4.107}$$

Assumption (3.4.105) bounds the conditional probabilities to see upward jumps of the particular sizes uniformly from below such that the probability for infinitely large jumps must be 0; the second assumption, (3.4.106), is the *usual* drift condition, which we already know from the last sections, in a stronger formulation.

Proof. We construct $(\hat{p}^-_m)_{m \in \{-1\} \cup \mathbb{N}}$ by

$$\hat{p}^-_{-1} := 1 - \delta, \tag{3.4.108}$$

$$\hat{p}^-_1 := 0, \tag{3.4.109}$$

$$\hat{p}^-_m := (1 - \hat{p}^-_{-1})(1 - q)^{m-2} q, \quad m \in \mathbb{N} \backslash \{1\}. \tag{3.4.110}$$

$(\hat{p}^-_m)$ is a probability distribution with

$$\hat{p}^-_{-1} + \sum_{k=1}^m \hat{p}^-_k \leq p^-_{n,-1}(e) + \sum_{k=1}^m p^-_{n,k}(e) \tag{3.4.111}$$

for all $e \in [0, C^n]$, $n \geq n^*$ and $m \in \mathbb{N}$.

Let $r^* := C^{n^*-1}$. We set $t_0^- := \nu_0^+(r^*)$, and assume that $D_{t_0^-} \in (C^{n^*-1}, C^{n^*}]$; else we adjust n^* such that the assumption holds. We denote the smallest power of C that is larger than $x > 0$ by

$$\lceil x \rceil := \min\{C^n : n \in \mathbb{Z}, C^n \geq x\}, \qquad (3.4.112)$$

and define recursively

$$t_k^- := \min\left\{t > t_{k-1}^- : D_t \in \left\{\lceil D_{t_{k-1}^-}\rceil/C\right\} \cup \left[C\lceil D_{t_{k-1}^-}\rceil, \infty\right)\right\} \qquad (3.4.113)$$

for $k \in \mathbb{N}$. The random process $\Xi^- = (\Xi_k^-)_{k \in \mathbb{N}_0}$, given by

$$\Xi_k^- := \frac{\ln D_{t_k^-}}{\ln C}, \qquad (3.4.114)$$

has state space $\mathbb{R}$, and $\Xi_0^- \leq n^*$. We introduce the Markov chain $\bar\Xi^- = (\bar\Xi_k^-)_{k \in \mathbb{N}_0}$ defined by $\bar\Xi_0^- := n^*$ and the transition probabilities

$$P\left(\bar\Xi_{k+1}^- = n + m \mid \bar\Xi_k^- = n\right) := \hat p_m^- \qquad (3.4.115)$$

for $m \in \{-1\} \cup \mathbb{N}$. By the definition of $(\hat p_m^-)$ and by (3.4.111), we can couple Ξ^- and $\bar\Xi^-$ such that

$$\Xi_k^- \leq \bar\Xi_k^- \qquad (3.4.116)$$

for all $k \in \{0, \ldots, k^*\}$ with $k^* \in \mathbb{N}$ being the smallest index with $\Xi_{k^*}^- \leq n^*$. We are done if we can show that k^* is almost surely finite; yet, by the coupling argument, it also sufficient to show that there is almost surely a finite $\bar k^* \in \mathbb{N}$ such that $\bar\Xi_{\bar k^*}^- \leq n^*$. We apply again Theorem 3.6 with $\mathcal{E}_0 := \{n^* - 1\}$ and $h(i) := i$. We have to show

$$\hat p_{-1}^-(n-1) + \sum_{m=1}^\infty (n+m)\hat p_m^- \leq n. \qquad (3.4.117)$$

By the definition of $(\hat p_m^-)$, the inequality easily transforms to

$$\sum_{m=0}^\infty (m+2)(1-q)^m q \leq \frac{1-\delta}{\delta}. \qquad (3.4.118)$$

Finally, we solve the geometric series explicitly and use $\delta \leq q(1+2q)$ to see that

3.4. AN ANALOGY FOR HEAVY-TAILED LÉVY PROCESSES

(3.4.118) holds if
$$\frac{1-q}{q} + 2 \leq \frac{1+q}{q}. \tag{3.4.119}$$

We conclude the proof by observing that both sides are equal. $\square$

Proving that condition (3.4.105) of Lemma 3.14 holds

For the proof of (3.4.105), let us assume that we would like to sample the path of a Lévy process L, $L(\omega)$, such that the particular realisation of D, $D(\omega)$, started in C^n jumps into $[C^{n+m}, C^{n+m+1})$, $m \in \mathbb{N}\setminus\{1\}$, when it leaves (C^{n-1}, C^{n+1}) for the first time. We could first sample the time when the particular jump happens. Let us denote this time by τ. Observe that τ is a stopping time, since D is adapted to the natural filtration of L. Next, we could consider the set of paths of Lévy processes, that are conditioned in such a way that the particular D_t remains in (C^{n-1}, C^{n+1}) for all times in the interval $[0, \tau(\omega))$. Sampling a path from this set would imply to sample $D_{\tau(\omega)-}$ and $E_{\tau(\omega)-}$. Finally, we had to sample $\Delta_{\tau(\omega)}$ with respect to $D_{\tau(\omega)-}(\omega)$ and $E_{\tau(\omega)-}(\omega)$ such that $D_{\tau(\omega)}(\omega)$ was in the particular interval. Of course, this little gedankenexperiment fails, as we do not know the distributions of the objects that we would have to sample. Nevertheless, we can make one crucial observation: the exact distribution of Δ_τ depends on $D_{\tau-}$ and $E_{\tau-}$ only. Thus, we can find a lower bound for $q_{n,m}^-$ by minimising the probability to jump into the particular interval over all possible positions of $D_{\tau-}$ and $E_{\tau-}$. Let us abbreviate (C^{n-1}, C^{n+1}) by $\bar{C}_n$ for reasons of space, then the idea expressed in mathematical terms yields

$$\inf_{e \in [0, C^n]} q_{n,m}^-(e) \tag{3.4.120}$$
$$\geq \inf_{d \in \bar{C}_n} \inf_{e \in [0,d]} P\Big(\Delta \in \big(d - e - C^{n+m+1}, d - e - C^{n+m}\big]$$
$$\cup \big[C^{n+m} - e, C^{n+m+1} - e\big) \,\Big|\, \tag{3.4.121}$$
$$\Delta \in \big(-\infty, d - e - C^{n+m}\big] \cup \big[C^{n+m} - e, \infty\big)\Big).$$

Term (3.4.121) equals

$$
\inf_{d \in \bar{C}_n} \inf_{e \in [0,d]} \left[\frac{P\left(\Delta \in \left(d - e - C^{n+m+1}, d - e - C^{n+m}\right]\right)}{P\left(\Delta \in (-\infty, d - e - C^{n+m}] \cup [C^{n+m} - e, \infty)\right)} \right. \tag{3.4.122}
$$
$$
\left. + \frac{P\left(\Delta \in [C^{n+m} - e, C^{n+m+1} - e)\right)}{P\left(\Delta \in (-\infty, d - e - C^{n+m}] \cup [C^{n+m} - e, \infty))} \right].
$$

$$
\geq \inf_{d \in (C^{n-1}, C^{n+1})} \inf_{e \in [0,d]} \frac{c^-}{c^+} \left[\frac{\int_{C^{n+m}-e}^{C^{n+m+1}-e} x^{-(\alpha+1)} dx}{\int_{C^{n+m}-e}^{\infty} x^{-(\alpha+1)} dx + \int_{C^{n+m}+e-d}^{\infty} x^{-(\alpha+1)} dx} \right. \tag{3.4.123}
$$
$$
\left. + \frac{\int_{C^{n+m}+e-d}^{C^{n+m+1}+e-d} x^{-(\alpha+1)} dx}{\int_{C^{n+m}-e}^{\infty} x^{-(\alpha+1)} dx + \int_{C^{n+m}+e-d}^{\infty} x^{-(\alpha+1)} dx} \right]
$$

$$
= \inf_{d \in (C^{n-1}, C^{n+1})} \inf_{e \in [0,d]} \frac{c^-}{c^+} \left[1 - \frac{\left(C^{n+m+1} - e\right)^{-\alpha}}{\left(C^{n+m} - e\right)^{-\alpha} + \left(C^{n+m} + e - d\right)^{-\alpha}} \right. \tag{3.4.124}
$$
$$
\left. + \frac{\left(C^{n+m+1} + e - d\right)^{-\alpha}}{\left(C^{n+m} - e\right)^{-\alpha} + \left(C^{n+m} + e - d\right)^{-\alpha}} \right]
$$

$$
= \inf_{\bar{d} \in (C^{-m-1}, C^{-m+1})} \frac{c^-}{c^+} \left[1 - \left(\frac{1 - \bar{d}/2}{C - \bar{d}/2} \right)^\alpha \right] \tag{3.4.125}
$$

with $\bar{d} = dC^{-(n+m+1)}$. The last expression is arbitrary close to c^-/c^+ for a sufficiently large constant C. Thus,

$$
\inf_{e \in [0, C^n]} q^-_{n,m}(e) \geq \frac{c^-}{c^+} - \epsilon \tag{3.4.126}
$$

for all $\epsilon > 0$.

Proving that condition (3.4.106) of Lemma 3.14 holds

We prove (3.4.106) by showing that $p^-_{n,-1}$ tends to 1. The procedure works exactly as in the Brownian case: We calculate how much time X and Y need to reach the boundaries of a tube with radius C^{n-2}. Let us denote this time by T^*_n. The probability $p^-_{n,-1}$ is greater than the probability that L remains in that tube until time T^*_n if started in the centre. For the detailed calculations, we refer to our statements from page 48, starting below the proof of Lemma 3.5, to page 49, ending at line (3.3.73). It follows

$$p^-_{n,-1} \geq P\left(\sup_{s\in[0,T^*_n]} |L_s| \leq C^{n-2}\right) \quad (3.4.127)$$

for

$$T^*_n = \frac{1}{\gamma+1}\left[\left(\frac{C^2+1}{C^2}C^n + K\right)^{\gamma+1} - \left(\frac{C^2-1}{C^2}C^n + K\right)^{\gamma+1}\right] \in \Theta\left(C^{n(\gamma+1)}\right). \quad (3.4.128)$$

There is no general scaling property for Lévy processes as we use it in the Brownian case. Furthermore, the processes considered here have no variance because of their large jumps. We solve this problem by introducing a new Lévy process $\hat{L}$ that does not have large jumps; thus a second moment exists and we can apply *Doob's L^p-inequality*. Additionally, there is a high probability that the removed large jumps do not appear in the considered time horizon anyway such that we can just substitute the original Lévy process by $\hat{L}$.

Recall that $L \in \mathfrak{C}(\alpha)$ with characteristic triplet $(\mu, 0, \lambda)$ is a martingale. We can choose a constant $c_n > 0$ such that $\hat{L}$ with characteristic triplet $(\mu, 0, \hat{\lambda})$,

$$\hat{\lambda}(dx) := \mathbb{1}_{\{x \in [-c_n C^n, C^n]\}} \lambda(dx), \quad (3.4.129)$$

is also a martingale; in words, if we cut off all positive jumps larger than C^n, we can compensate the cut by removing all negative jumps smaller than some value $-c_n C^n$. Since $\lambda((-\infty, -x])$ and $\lambda([x, \infty))$ are of the same order as x tends to infinity, C^n and $c_n C^n$ must also be of the same order for large n.

Let $(\Delta_t)_{t\geq 0}$ be the jump size of L at time t,

$$\Delta_t := L_t - L_{t-}, \quad (3.4.130)$$

and

$$T_{x,y} := \min\{t \geq 0 : (\Delta_t < x) \vee (\Delta_t > y)\} \quad (3.4.131)$$

the first time that L makes a jump smaller than $x < 0$ or larger than $y > 0$. We

have
$$c^+ \int_{C^n}^{\infty} x^{-(\alpha+1)} dx + c^+ \int_{-\infty}^{c_n C^n} x^{-(\alpha+1)} dx \in \Theta\left(C^{-n\alpha}\right) \qquad (3.4.132)$$

for $c^+ > 0$, and thus

$$P\left(T_{-c_n C^n, C^n} > T_n^*\right) \in \Theta\left(\exp\left(-C^{n(\gamma+1-\alpha)}\right)\right), \qquad (3.4.133)$$

which tends to 1 for $\gamma < \alpha - 1$.

We continue line (3.4.127) and write

$$p_{n,-1}^- \geq P\left(T_{-c_n C^n, C^n} > T_n^*\right)$$

$$P\left(\sup_{s \in [0, T_n^*]} |L_s| \leq C^{n-2} \;\middle|\; T_{-c_n C^n, C^n} > T_n^*\right) \qquad (3.4.134)$$

$$= P\left(T_{-c_n C^n, C^n} > T_n^*\right) P\left(\sup_{s \in [0, T_n^*]} |\hat{L}_s| \leq C^{n-2}\right) \qquad (3.4.135)$$

$$\in \Theta\left(\exp\left(-C^{n(\gamma+1-\alpha)}\right) C^{n((2-\alpha)+(\gamma+1)-2)}\right). \qquad (3.4.136)$$

In the last line, we used again Doob's L^p inequality (see Theorem 3.13) for $p = 2$ and the statement that

$$\mathbb{E}(\hat{L}_t^2) \in \Theta\left(tC^{n(2-\alpha)}\right). \qquad (3.4.137)$$

We have already calculated the second moment of a *cropped* Lévy process in the lines (3.4.87) to (3.4.90). As the calculation here is similar, we do not go into detail again. To finish the proof of (3.4.106), observe that

$$(n((2-\alpha) + (\gamma+1) - 2) < 0) \Leftrightarrow (\gamma < \alpha - 1). \qquad (3.4.138)$$

Chapter 4

Optimal execution strategies for large orders

4.1 Introduction

> *Mon dessin ne représentait pas un chapeu.*
> *Il représentait un serpent boa qui digérait un éléphant.*
> Antoine de Saint-Exupery - Le petit prince

We have already mentioned in Section 1.1 that option pricing is one of the main issues in financial mathematics, but the scope has been extended in the last years. One of the current topics of interest is the theory of optimal trading strategies for the execution of large orders. Here, a trader would like to purchase[a] a huge volume of shares up to time T. Since the supply of limit orders for a certain price is limited, the trader will not be able to trade the whole order for the current price, but he or she will suffer from an adverse price movement. This additional price impact, induced by the trader's own trading, can be lessened if he or she gives the market time to recover; the *best price* returns to previous levels. However, the time interval $[0, T]$ is assumed to be too short in order to wait for a full recovery of the market. The *optimal execution problem* asks for the optimal splitting and the optimal trading times to minimise the expected price impact.

There have been several models to solve the optimal execution problem, moti-

---

[a]In this chapter, we focus on a trader *purchasing* shares, since all models mentioned here either consider only this part of the problem or make symmetric assumptions for buyers and sellers.

vated by empirical findings (for references see next paragraph); yet, since we do not know if these models capture all relevant features of real markets, we cannot be sure that the strategies work in reality, and tests on real markets would be an expensive experiment. For this reason, the Opinion Game is an excellent tool for testing theoretical models of optimal trading strategies; it provides an artificial, yet reasonable, market environment that allows for applying the strategies without costs or risk, comparing the numerical results with the theoretical expectations and resolving deviations by an improvement of the underlying market assumptions with respect to the empirical findings. Furthermore, the Opinion Game provides a more realistic market response to orders than classical order book models, since its generalised order book also captures traders who are willing to trade for a price close to the best quotes but have not placed public orders (in some markets it is also possible to place hidden or partially visible orders (Frey and Sandas (2009); Bessembinder et al. (2009))). These traders offer *hidden liquidity*; they will influence the price impact when an order is executed but do not appear in the order book (Weber and Rosenow (2005)). In this chapter, we exemplify in detail how the Opinion Game can contribute to *better* solutions for the optimal execution problem.

All approaches to the optimal execution problem rely on two empirical findings that have been validated in many studies (see Schöneborn (2008), page 3, for a list of references): First, a large order has an impact on its price; second, this impact decreases in time, but it does not vanish completely. That implies the costs of all subsequent orders are influenced by the impact of a large order. These two effects are called *temporary* and *permanent impact*. Many models implement these observations straightly (Bertsimas and Lo (1998); Almgren and Chriss (2001); Huberman and Stanzl (2005)): They consider a stochastic process that simulates the current best price evolving independently from the large trader's action in time, and two functions mapping the volume of a large order to the temporary or, respectively, permanent impact. When a large order is executed, its temporary impact and the previous orders' permanent impacts are just added to the price.

4.1. INTRODUCTION

Yet, it is doubtful if the complex dynamics of limit order books (LOB), which underlie most modern markets, can be captured by looking at the best price only. Therefore, recent models attempt to take the dynamics of the whole order book into account. Obizhaeva and Wang introduced a model with an underlying block-shaped LOB and calculated the optimal trading strategy in terms of a recursive formula by applying Bellman equations (Obizhaeva and Wang (2005)). Alfonsi, Fruth and Schied introduced a generalisation of this model for general order book shapes and gave an explicit solution for the optimal trading strategy with respect to their market model (introduced in Alfonsi et al. (2010) and revisited in Alfonsi and Schied (2009)); this model is the one we will test in a microscopic market environment, and we refer to it as the *AFS model*.

The AFS model describes the underlying market by two parameters: The shape of the LOB given by a *shape function f* and a positive constant ρ controlling the *resilience speed* of the order book. There are two versions of the model: In the first one, the consumed volume recovers exponentially fast; in the second version, the best price recovers in this way. The shape of the order book is *static* such that there is a bijection between the impact on the best price and on the volume. Thus, the response of the order book to the execution of a large order depends on the current price impact only, but not on possible executions before.

To test the AFS model, we first have to determine the correct values for f and ρ in the Opinion Game. There are several problems to find the value for ρ. First, the AFS model does not assume a permanent impact in the assumptions; second, the market recovery is only poorly approximated by an exponential function; third, ρ does not exist as a constant value but depends on the traded volume. While the first two items can be bypassed, the third item strongly conflicts with the assumptions of the AFS model and leads to the main result of this chapter.

We introduce a generalisation of the AFS model that we call the *generalised AFS model* or *GAFS model*. The GAFS model substitutes ρ by $\bar{\rho}$ that is a function of an order's price impact or volume impact, depending on the model version. We prove that there exists a unique, deterministic optimal trading strategy for the

GAFS model. This optimal strategy has an interesting feature: To determine it, we have to compute the unique root r of a certain function that depends on f and $\bar{\rho}$. The optimal strategies for the AFS and the GAFS model coincide if we set $\rho := \bar{\rho}(r)$. In this sense, the GAFS optimal strategy finds the *correct* ρ value for the AFS model.

We support our analytical findings by simulations in the Opinion Game. We compute the GAFS optimal strategies for several parameter sets and apply them on the Opinion Game market. It turns out that, although the AFS strategy with respect to the *correct* value for ρ coincides with the GAFS optimal strategy, a *bad*, yet reasonable, choice of the value for ρ produces significantly higher sampled impact costs; a justification for the GAFS model's benefit. Nevertheless, also the GAFS model does still not capture all relevant details of the Opinion Game's order book dynamics. The sampled costs may show the expected qualitative behaviour, for example they decrease if the available trading time T or the number of trading opportunities becomes larger, but, in comparison to the predicted costs, they are up to four times higher; the predictions of the GAFS model fail completely on a quantitative level. We give a heuristic explanation for this deviation, suggesting possible ways to make the (G)AFS market model more realistic.

Let us shortly summarise the structure of this chapter. In Section 4.2, we introduce the AFS model in detail and state its optimal trading strategies. In Section 4.3, we extend the algorithm of the Opinion Game to large orders. In Section 4.4, we try to determine f and ρ in the Opinion Game; the problems thereby lead to the GAFS model and its optimal strategies. In Section 4.5, we apply the GAFS optimal strategies in the Opinion Game, and demonstrate that they perform better than the AFS strategies. Furthermore, we compare the costs generated by the GAFS strategies in simulations to the theoretically predicted costs. There are significant deviations. In Section 4.6, we give some ideas where the deviations originate from. These ideas can serve as a basis for improving the (G)AFS market model. Finally, we prove the GAFS optimal strategies from Section 4.4 in Section 4.7.

4.2 The AFS market model, Version 1 & 2

A trader would like to purchase $X_0 > 0$ shares within a time period $[0, T]$, $T > 0$. X_0 is assumed to be large such that the trader's order has an impact on the price and the underlying limit order book. We refer to this trader as *large trader* in the following. Because we consider a buy order, we first define how the *upper part* of the LOB, which contains the sell limit orders, is modelled. As long as the large trader does not take action, the LOB is described by the *unaffected best ask price* $A^0 = (A_t^0)_{t \geq 0}$ and by the *shape function* $f : \mathbb{R} \to (0, \infty)$. A^0 is a martingale on a given filtered probability space $(\Omega, (\mathcal{F}_t)_{t \geq 0}, \mathcal{F}, P)$ satisfying $A_0^0 = A_0$ for some $A_0 \in \mathbb{R}$; f is a continuous function. The amount of shares available for a price $A_t^0 + x$, $x \geq 0$, at time t is then given by $f(x)dx$. Notice that the shape of the order book with respect to the best ask price is static.

Now, assume the large trader acts for the first time and purchases x_0 shares at time t_0; he or she consumes all shares between $A_{t_0}^0$ and $A_{t_0}^0 + D_{t_0+}^A$, $D_{t_0+}^A$ being uniquely determined by

$$\int_0^{D_{t_0+}^A} f(x)dx = x_0. \tag{4.2.1}$$

This formula results from the assumption that the LOB is block-shaped with height 1. $D^A = (D^A)_{t \geq 0}$ is called the *extra spread* caused by the large trader. In general, if we know $D_{t_n}^A$ for a trading time t_n, $D_{t_n+}^A$ is given by

$$\int_{D_{t_n}^A}^{D_{t_n+}^A} f(x)dx = x_n \tag{4.2.2}$$

whereby x_n is the amount of shares traded at time t_n. The large trader is inactive between two trading times, t_n and t_{n+1}, and the extra spread recovers. For the exact way of recovery there are two versions considered. To conform to the notation of Alfonsi et al. (2010), we first state *Version 2*. In this case, D_t^A is defined for $t \in (t_n, t_{n+1}]$ by

$$D_t^A := e^{-\rho(t-t_n)} D_{t_n+}^A. \tag{4.2.3}$$

The parameter ρ is a positive constant called the *resilience speed*. To complete the definition, we set $D_t^A := 0$ for $t \leq t_0$. Now, we can introduce the *best ask price*

$A = (A_t)_{t \geq 0}$ by

$$A_t := A_t^0 + D_t^A. \qquad (4.2.4)$$

In contrast to A^0, A includes the large trader's impact. In particular, the amount of shares available for a price $A_t^0 + x$ at time t is given by

$$\begin{cases} f(x)dx & \text{for } x \geq A_t - A_t^0 \\ 0 & \text{otherwise} \end{cases}. \qquad (4.2.5)$$

In other words, every trader in the market experiences the large trader's impact after time t_0.

The *price impact* D^A can also be expressed in terms of the *impact on the volume* $E^A = (E_t^A)_{t \geq 0}$. Because the shape function f is strictly positive, there is a one-to-one relation between E^A and D^A. Given D^A, the process E^A is defined by

$$E_t^A := \int_0^{D_t^A} f(x)dx. \qquad (4.2.6)$$

We introduce the antiderivative of f,

$$F(x) := \int_0^x f(x)dx, \qquad (4.2.7)$$

to get the relations

$$E_t^A = F(D_t^A) \quad \text{and} \quad D_t^A = F^{-1}(E_t^A). \qquad (4.2.8)$$

By (4.2.2) and (4.2.8), we easily conclude

$$E_{t_n+}^A = E_{t_n}^A + x_n. \qquad (4.2.9)$$

This motivates to define *Version 1*, in which we first define E^A and then derive D^A by relation (4.2.8). We set $E_t^A := 0$ for $t \in [0, t_0]$ and

$$E_t^A := e^{-\rho(t-t_n)} E_{t_n+}^A, \quad t \in (t_n, t_{n+1}]. \qquad (4.2.10)$$

The equations (4.2.9) and (4.2.10) define E^A completely.

Summarising, we have introduced two versions of the AFS model: In Version 1, we define the volume impact E^A and assume that it recovers exponentially fast between the large trader's orders. D^A is then derived from E^A by relation (4.2.8). In Version 2, we first define the price impact D^A, assume an exponentially fast recovery and derive E^A from it.

We cannot exclude a priori that it is reasonable to sell shares and to buy them back later. Thus, we also have to model the impact of (large) sell orders on the LOB. Such orders will be written as orders with negative sign. Let $B^0 = (B^0_t)_{t \geq 0}$ be the *unaffected best bid price* with

$$B^0_t \leq A^0_t \text{ for all } t \geq 0 \qquad (4.2.11)$$

as only constraint for its dynamics. The *lower part* of the LOB is modelled by the shape function f on the negative part of its domain. More precisely, the number of bids for the price $B^0_t + x$, $x < 0$, is given by $f(x)dx$. As before, we can now introduce the *extra spread* $D^B = (D^B_t)_{t \geq 0}$. Given a sell order $x_n < 0$, a trading time t_n and the particular extra spread $D^B_{t_n}$, $D^B_{t_n+}$ is implicitly defined by

$$\int_{D^B_{t_n}}^{D^B_{t_n+}} f(x)dx = x_n. \qquad (4.2.12)$$

Note that D^B is nonpositive. We define the *impact on the volume* $E^B = (E^B_t)_{t \geq 0}$ by

$$E^B_{t_n+} := E^B_{t_n} + x_n. \qquad (4.2.13)$$

E^B is also nonpositive, and its connection to D^B is again given by (4.2.8). To complete the definitions for sell orders, we set $D^B_t := 0$ and $E^B_t := 0$ for all $t \leq t_0$, and

$$\begin{cases} E^B_t := e^{-\rho(t-t_n)} E^B_{t_n+} & \text{for Version 1} \\ & \text{for } t \in (t_n, t_{n+1}], \\ D^B_t := e^{-\rho(t-t_n)} D^B_{t_n+} & \text{for Version 2} \end{cases} \qquad (4.2.14)$$

whereby t_n and t_{n+1} are two successive trading times of the large trader.

Now that all orders are well-defined, we introduce the *cost* of a large order x_{t_n}

at some trading time t_n by

$$\pi_{t_n}(x_{t_n}) := \begin{cases} \int_{D^A_{t_n}}^{D^A_{t_n}+} (A^0_{t_n} + x) f(x) dx & \text{for a buy market order } x_{t_n} \geq 0 \\ \int_{D^B_{t_n}}^{D^B_{t_n}+} (B^0_{t_n} + x) f(x) dx & \text{for a sell market order } x_{t_n} < 0 \end{cases}. \quad (4.2.15)$$

We assume that the large trader needs to purchase the X_0 shares in $N+1$ steps at equidistant points in time $0 =: t_0 < \ldots < t_N := T$. His or her *admissible strategies* are sequences $\xi = (\xi_0, \ldots, \xi_N)$ of random variables such that

- $\sum_{n=0}^N \xi_n = X_0$,
- ξ_n is $\mathscr{F}_{t_n}$-measurable for all n, and
- all ξ_n are bounded from below.

We denote the set of all admissible strategies by $\hat{\Xi}$. The goal is to find an admissible strategy ξ^* that minimises the *average cost* $\mathscr{C}(\xi)$ given by the the mean of the sum of the single trades' costs:

$$\mathscr{C}(\xi) := \mathbb{E}\left(\sum_{n=0}^N \pi_{t_n}(\xi_n)\right). \quad (4.2.16)$$

Under the technical assumption that

$$\lim_{x \to \infty} F(x) = \infty \quad \text{and} \quad \lim_{x \to -\infty} F(x) = -\infty, \quad (4.2.17)$$

Alfonsi, Fruth and Schied give the unique optimal strategies for both versions explicitly. For the sake of convenience, we set $\tau := T/(N+1) = t_{n+1} - t_n$.

Theorem 4.1 (Optimal strategy for Version 1, Theorem 4.1 in Alfonsi et al. (2010))**.** *Suppose that the function*

$$h_1(x) := F^{-1}(x) - e^{-\rho\tau} F^{-1}(e^{-\rho\tau} x) \quad (4.2.18)$$

is one-to-one. Then there exists a unique optimal strategy $\xi^{(1)} = (\xi_0^{(1)}, \ldots, \xi_N^{(1)})$. *The initial market order* $\xi_0^{(1)}$ *is the unique solution of the equation*

$$F^{-1}\left(X_0 - N\xi_0^{(1)}(1 - e^{-\rho\tau})\right) = \frac{h_1(\xi_0^{(1)})}{1 - e^{-\rho\tau}}, \quad (4.2.19)$$

the intermediate orders are given by

$$\xi_1^{(1)} = \cdots = \xi_{N-1}^{(1)} = \xi_0^{(1)}(1 - e^{-\rho\tau}), \qquad (4.2.20)$$

and the final order is determined by

$$\xi_N^{(1)} = X_0 - \sum_{n=0}^{N-1} \xi_n^{(1)}. \qquad (4.2.21)$$

In particular, the optimal strategy is deterministic. Moreover, it consists only of nontrivial buy orders, that is $\xi_n > 0$ for all n.

Theorem 4.2 (Optimal strategy for Version 2, Theorem 5.1 in Alfonsi et al. (2010)). *Suppose that the function*

$$h_2(x) := x \frac{f(x) - e^{-2\rho\tau} f(e^{-\rho\tau} x)}{f(x) - e^{-\rho\tau} f(e^{-\rho\tau} x)} \qquad (4.2.22)$$

is one-to-one and that the shape function satisfies

$$\lim_{|x|\to\infty} x^2 \inf_{y \in [e^{-\rho\tau}x, x]} f(y) = \infty. \qquad (4.2.23)$$

Then there exists a unique optimal strategy $\xi^{(2)} = (\xi_0^{(2)}, \ldots, \xi_N^{(2)})$. *The initial market order* $\xi_0^{(2)}$ *is the unique solution of the equation*

$$F^{-1}\left(X_0 - N[\xi_0^{(2)} - F(e^{-\rho\tau} F^{-1}(\xi_0^{(2)}))]\right) = h(F^{-1}(\xi_0^{(2)})), \qquad (4.2.24)$$

the intermediate orders are given by

$$\xi_1^{(2)} = \cdots = \xi_{N-1}^{(2)} = \xi_0^{(2)} - F(e^{-\rho\tau} F^{-1}(\xi_0^{(2)})), \qquad (4.2.25)$$

and the final order is determined by

$$\xi_N^{(2)} = X_0 - \sum_{n=0}^{N-1} \xi_n^{(2)}. \qquad (4.2.26)$$

In particular, the optimal strategy is deterministic. Moreover, it consists only of nontrivial buy orders, that is $\xi_n > 0$ for all n.

One can easily check that the orders $\xi_1^{(\cdot)}, \ldots, \xi_{N-1}^{(\cdot)}$ have exactly the volume that has recovered since the last trade. In this sense, the theorems just give the right balance between the first and the last order. This balance is found by solving the particular equations, (4.2.19) and (4.2.24), given in both theorems.

4.3 The Opinion Game extended to large orders

Before we can test the AFS strategies, we have to explain how large orders are defined in the Opinion Game. The first problem is already that there is no explicit notion of orders; consequently, also large orders and their executions are not defined a priori. Yet, there must be the notion of orders on an implicit level, since the Opinion Game simulates a double auction market. On such markets, a trade can only happen if a buy and a sell order match. Consequently, the Opinion Game makes the underlying assumption that some opinions are public, that is they are quoted as orders, at least the ones of those agents that are trading. This observation motivates a change of our perspective on the Opinion Game: In the remainder of this chapter, we rather think about (maybe hidden or unplaced) buy or sell orders instead of traders with opinions[b], and omit again the word "generalised" in the following when we talk about the (generalised) order book of the Opinion Game.

The existence of orders is the basis to define large orders and the corresponding execution procedure. We use the notation of the Opinion Game's algorithm, introduced in Chapter 2, and assume we would like to purchase X stocks at time t. Then, we do not apply the standard dynamics at time t, but we use the following algorithm:

set $p_k^{(1)} := p_k(t)$ for all $k \in \{1, \ldots, N\}$

set $n_k^{(1)} := n_k(t)$ for all $k \in \{1, \ldots, N\}$

let $p^a(1)$ be the best ask price of the configuration $\left(p_k^{(1)}, n_k^{(1)}\right)_{k \in \{1, \ldots, N\}}$

from $x := 1$ to X do {

---

[b]Recall that we have already argued in Chapter 2 that it can be favourable to consider single shares or demands instead of agents.

find i s.th. $p_i^{(x)} \leq p_j^{(x)}$ for all $j \in \{1, \ldots, N\}$

$p_i^{(x+1)} := p^a(x)$

choose uniformly trading partner j s.th. $i \neq j$ and $p_j^{(x)} = p^a(x)$

$n_i^{(x+1)} := 1$ and $n_j^{(x+1)} := 0$

$p_j^{(x+1)} := p^a(x) - g$

$p_i^{(x+1)} := p^a(x) + \hat{g}(x)$

set $p_k^{(x+1)} := p_k^{(x+1)}$ for all $k \in \{1, \ldots, N\} \setminus \{i, j\}$

set $n_k^{(x+1)} := n_k^{(x+1)}$ for all $k \in \{1, \ldots, N\} \setminus \{i, j\}$

let $p^a(x+1)$ be the best ask price of $\left(p_i^{(x+1)}, n_i^{(x+1)}\right)_{i \in \{1, \ldots, N\}}$

}

set $p_k(t+1) := p_k^{(X+1)}$ for all $k \in \{1, \ldots, N\} \setminus \{i, j\}$

set $n_k(t+1) := n_k^{(X+1)}$ for all $k \in \{1, \ldots, N\} \setminus \{i, j\}$

The value g is the same random or deterministic value as in the original dynamics. The random variables $\hat{g}(x)$ are independently distributed with measure

$$P(\hat{g}(x) = k) = \frac{1}{M} \sum_{n=1}^{N} \mathbb{1}_{\{p_n^{(x)} - p^a(x) = k\}} \text{ for } k \in \mathbb{N}_0. \tag{4.3.1}$$

In other words, we execute a large buy order of volume X by taking the lowest X orders one by one and putting them directly to the ask price such that a trade is enforced. The number of market participants is constant in the Opinion Game, thus taking orders from the tail is an obvious method to simulate a large order that is placed *out of the blue*. After each single trade, we adjust the order prices; the price of the (new) buy order is decreased by g, the price of the sell order is increased by $\hat{g}$. The density function of $\hat{g}$ is given by the order book's current shape.

This choice of $\hat{g}$ leads to a realistic response of the order book to the execution of large orders (see Figure 4.1). While the large order is executed, the new sell orders have a great probability to be placed in vicinity to the peak of the order book's

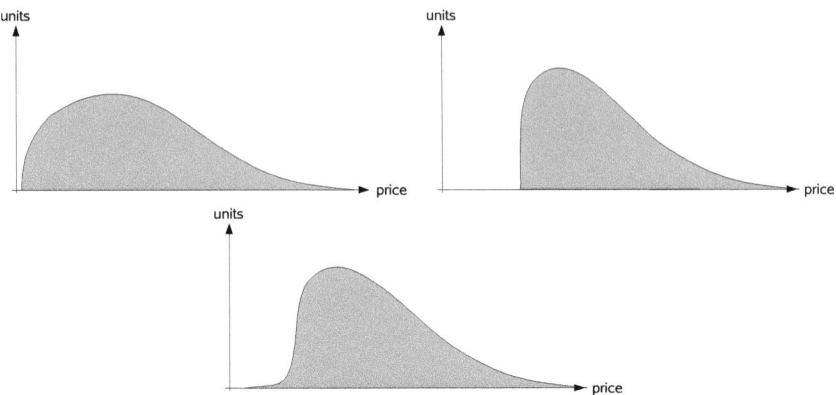

Figure 4.1: Sketch of the order book shape in the Opinion Game when a large order is executed. Before the execution, the order book is in equilibrium (upper left figure); directly afterwards, the best ask price is increased, and there is more liquidity close to it (upper right figure). When the LOB recovers from the order, the best ask price decreases, but the best quotes have a low volume only (lower figure); it takes more time until the order book is in equilibrium again.

seller part; thus the peak grows, and the order book provides more liquidity for prices in this region. Here, we implement the idea that the execution of a large buy order leads to a conspicuous rise in the price that attracts more traders to place sell orders close to the current best ask price; these traders hope that the price increase continues such that their orders are executed. At the same time, these additional offers provide more liquidity that slows down the price increase. If we consider the immediate price impact of the large order as function of the executed volume, the additional liquidity leads to a sublinear function shape. Sublinear behaviour of an order's immediate price impact has also been observed for real world markets in several empirical studies (Bouchaud et al. (2004); Almgren et al. (2005)). After the execution of the large order, the price increase stops and some traders realise quickly that orders for higher prices will probably not be executed in the near future; they place new orders for lower prices. However, most traders need more time to acknowledge that their price claims may be too high. In result, the best ask price decreases, but the order book volume in proximity to the new best quote is low. It takes more time until the LOB is back in equilibrium. This recovery behaviour of the order book is technically implemented by the preference for traders close to the best quotes in equation (2.2.2) when we update opinions.

As another feature that is known from real world markets, the best ask price does not return to the value it has had before the execution, but it stabilizes at higher values after the order book has returned to equilibrium. We discuss this *permanent impact* on the best price in Section 4.4.2.

Since the dynamics are symmetric, the algorithm applies to large sell orders in the same way.

4.4 GAFS - a generalisation of the AFS model

The actual goal of this section is determining the values of f and ρ for the Opinion Game. While this is a solvable task for f, which we deal with in Section 4.4.1, it is impossible to determine ρ. In particular, it turns out that the assumption of a constant ρ is not valid in the Opinion Game. A sophisticated procedure is needed to bypass this problem. We first substitute ρ by a function $\bar{\rho}$ that maps both the order's impact and the time elapsed since the last trade to the resilience speed. Then, we describe how we can extract the function values from the sampled data, and argue that it is sufficient to know the impact-dependent function $\bar{\rho}(\cdot) := \bar{\rho}(\cdot, \tau)$ only; recall that $\tau = T/N$ was the recovery time between two successive trades. The final problem, $\bar{\rho}$ is still depending on the order's impact, motivates to define the GAFS market model that differs only in one point from the AFS model: instead of a constant resilience speed ρ, there is the function $\bar{\rho}$ that maps the price impact (in Version 2, Section 4.4.2) or the volume impact (in Version 1, Section 4.4.3) to the resilience speed. The highlights of this chapter are the Theorems 4.3 and 4.4 that state the optimal trading strategies for the GAFS model.

For all simulations of the Opinion Game that we did for this chapter, we used the parameter values of Table 2.1. The strength of the external signal, δ_{ext}, was 1 for all times, as we wanted to focus on the intrinsic behaviour of the market; the external signal would have blurred our observations. Although a further variation of parameters surely leads to additional insight, our choice already gives a sound understanding of the problems that occur when applying the AFS model.

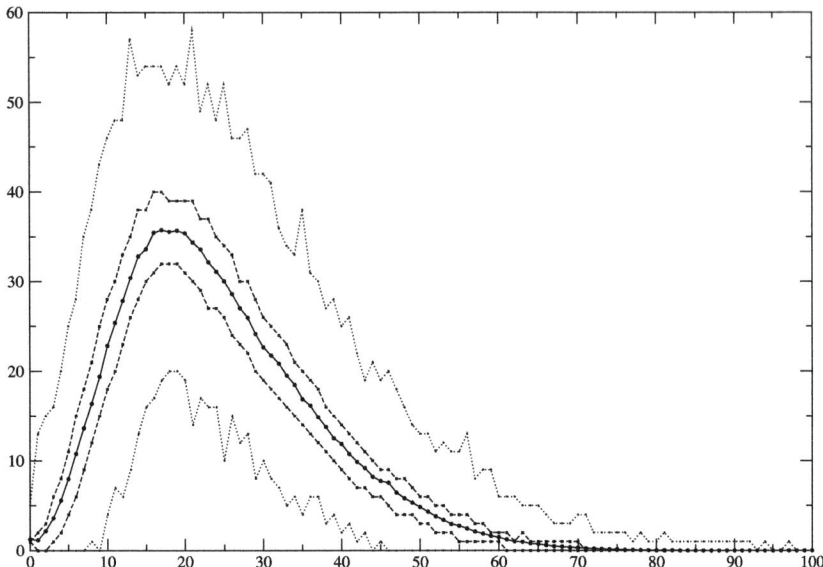

Figure 4.2: The seller part of the LOB relative to the best ask price. The solid line marks the mean values, the dashed lines illustrate the quartiles. The minimal and maximal values are illustrated by the dotted lines.

All sample runs were started independently with a new instance of the random number generator. Furthermore, the recording of data or the execution of large orders was started after 1 000 000 simulation steps only such that the model had sufficient time to get close to a stable state.

4.4.1 The shape of the Opinion Game's order book

To determine f, we recorded 500 times the Opinion Game's LOB relative to the best prices. Figure 4.2 shows the resulting upper part of the order book. The lower part is symmetric up to small deviations caused by the object's random nature. Even if the shape is not static as assumed in the AFS model, an *averaged shape* is clearly visible. We use these mean values to define the shape function f for the Opinion Game. For non-integer values, we interpolate f by assuming that the function is a right-continuous step function. This means that we violate the assumption of the AFS model about f being continuous. Yet, this choice for f has the advantage that the integral of f from 0 to an integer n is equal to the sum of the integer function values from 0 to $n-1$. Furthermore, for all parameter sets that we considered, we were still able to find unique solutions for the optimal trading strategies.

Recall that the price scale in the Opinion Game is logarithmic, whereas the AFS model assumes a linear scale. However, we can scale the grid of the Opinion Game with the parameter ϵ; the difference between relative and absolute price changes is negligible if ϵ is small. To determine the order of ϵ, we consider an order of 200 units of shares, 20% of the market volume in the Opinion Game; it is mentioned in Alfonsi et al. (2010) that the size of large orders can amount up to 20% of the daily traded volume. We assume that the shape of the LOB, f, is determined as described above, and the best ask price before our trade is denoted by A^0. Then, the relative impact costs are given by

$$\frac{1}{200 e^{\epsilon A^0}} \int_0^{D_{0+}} e^{\epsilon(A^0 + x)} f(x) dx - 1 \approx \frac{\epsilon}{200} \underbrace{\int_0^{D_{0+}} x f(x) dx}_{\approx 2039.47} \approx 10.20\epsilon. \qquad (4.4.1)$$

An empirical study at the US stock market shows that large orders can cause relative costs up to 3.55% (Almgren et al. (2005)). If we assume that $\epsilon \leq 0.0355/10.2$, ϵ is of order 10^{-3}. Even if our estimation of ϵ is rough, it nevertheless shows that a small value for ϵ is a reasonable assumption. However, we are interested in qualitative results; thus, and for the sake of convenince, we simply assume that the Opinion Game operates on $\mathbb{Z}$.

4.4.2 From determining ρ to the GAFS model, Version 2

In the following, we describe our approach to calibrate ρ for the Opinion Game. Recall that, in Version 2, this parameter determines the recovery speed of the price impact. We first describe how we sampled the necessary data. Afterwards, we focus on the main problems of extracting ρ from those data. Possible solutions are discussed and culminate in this section's main result: The GAFS theorem for Version 2, which assumes that the resilience speed is a function $\bar{\rho}$ depending on the order's price impact.

We fixed a price impact $D \in \{1, \ldots, 20\}$ and run 2500 simulations for each value of D. Each run consisted of a trading part in which a large sell order was executed at once. The particular order's volume was determined by its price impact: The trading part was finished as soon as the impact was equal to D. In a second experiment's part, we recorded the relaxation of the price. In particular, the large execution took place at time $\bar{t} := 1\,000\,000$; we recorded

$$\bar{p}(t) := p^a(t+1+\bar{t}) - p^a(\bar{t}) \tag{4.4.2}$$

for $t \in \{0, \ldots, 50\,000\}$, p^a denoting the best ask price of the Opinion Game order book (see Chapter 2). The process $(\bar{p}(t))_{t \in \mathbb{N}_0}$ is the discrete counterpart of the AFS model's process D^A.

To avoid problems caused by random fluctuations in $\bar{p}$, we consider the pointwise average of the samples denoted by $\langle \bar{p} \rangle$ and defined by

$$\langle \bar{p} \rangle_t := \frac{1}{2500} \sum_{i=0}^{2500} \bar{p}_t^i \tag{4.4.3}$$

for all $t \in \{0, \ldots, 50\,000\}$, $\bar{p}^i$ denoting the ith sample. For a clear distinction, we denote the value for ρ that we extract from $\langle \bar{p} \rangle$ by $\bar{\rho}_{\text{num}}$. The AFS model assumes $\langle \bar{p} \rangle$ to be of the form

$$\langle \bar{p} \rangle_t = D e^{-\bar{\rho}_{\text{num}} t} \tag{4.4.4}$$

with a static value $\bar{\rho}_{\text{num}}$; this follows from equation (4.2.3). Thus we should be able to determine $\bar{\rho}_{\text{num}}$ by

$$\bar{\rho}_{\text{num}} = \frac{\ln D - \ln \langle \bar{p} \rangle_t}{t} \tag{4.4.5}$$

for an arbitrary t. However, the right-hand side of the equation depends on D and t; thus, we would like to consider $\bar{\rho}_{\text{num}}(D, t)$ as a function.

Given $\langle \bar{p} \rangle$, let $\hat{p} : [0, \infty) \to \mathbb{R}$ the corresponding regression function of the form

$$\hat{p}_t := A + B e^{-\hat{\rho} t}, \tag{4.4.6}$$

for $t \in [0, \infty)$. It is determined by a Newton-Gauß algorithm with three degrees of freedom: A, B, $\hat{\rho}$. Observe that all three values can depend on D. The form of the regression function is motivated by assumption (4.4.4), which also leads to the expectation that $A = 0$ and $B = D$. Figure 4.3 shows the statistical behaviour of $\bar{p}$ for $D = 8$, the corresponding $\langle \bar{p} \rangle$ and $\hat{p}$. Furthermore, we compare $\langle \bar{p} \rangle$ for different D values. The three main problems are visible:

1. The AFS model assumes A_D to be 0; this is not the case.

2. The measured data is only well-approximated by an exponential function for large times. For small t, it is doubtful that the assumption of an exponential decay is the right choice at all.

3. If $\bar{\rho}_{\text{num}}$ was constant, the $\langle \bar{p} \rangle$ should be approximately parallel on a logarithmic scale; instead, $\bar{\rho}_{\text{num}}$ depends on D.

These problems occurred for all tested values of D. Next, we discuss the problems and their consequence for determining $\bar{\rho}_{\text{num}}$ one by one.

Existence of a permanent price impact

The existence of a permanent impact after the execution of a large order is well known. After having recovered, the LOB is shifted by $I_{\text{per}}(X)$, whereby $I_{\text{per}} : \mathbb{R} \to \mathbb{R}$ is assumed to be increasing, $I_{\text{per}}(0) = 0$ and negative arguments represent sell orders. Huberman and Stanzl (2004) argued on a theoretic level that linearity of I_{per} is equivalent to the absence of arbitrage opportunities. Empirical studies by Almgren et al. (2005) reinforce the conjecture of a linear permanent impact. Figure 4.4 shows the permanent impact for the Opinion Game. The mean is well-approximated by a linear function with coefficient 0.027.

Concerning the problems in determining $\bar{\rho}_{\text{num}}$, caused by the positive A_D, we have two possibilities: First, we could ignore the permanent impact such that $\bar{\rho}_{\text{num}}$ would be given by (4.4.5). This would be an appropriate solution for small t, but it would cause the AFS model to assume that even for large t the LOB is still not close to equilibrium; $\bar{\rho}_{\text{num}}$ could become arbitrarily small. Second, we could assume that the whole model has been shifted by A_D such that A_D is the

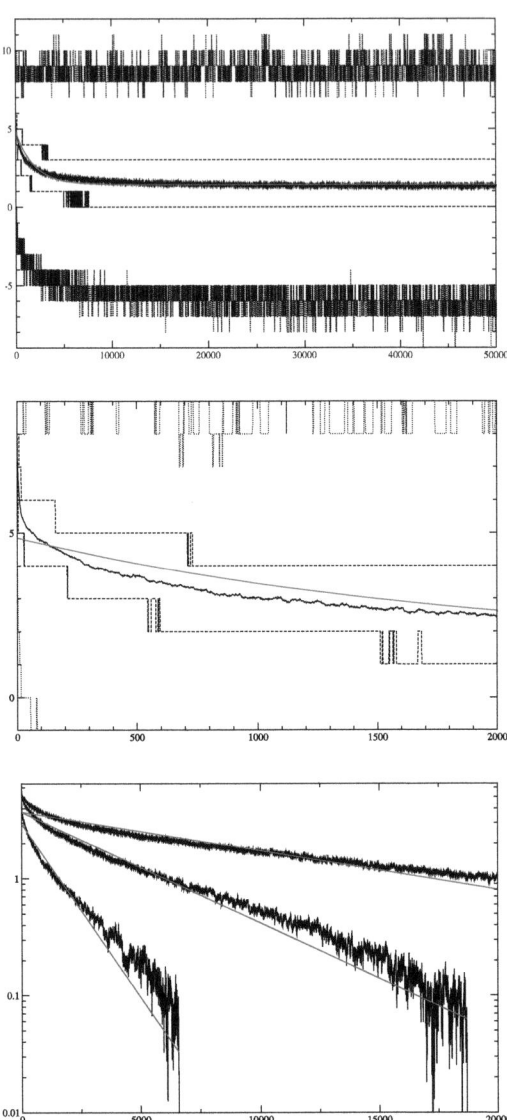

Figure 4.3: The two upper graphs show quartiles and extremal values of 2500 samples of $\bar{p}$ for $D = 8$, and the corresponding $\langle \bar{p} \rangle$ and $\hat{p}$ (grey). The upper graph illustrates the long-term behaviour on the domain $t \in [0, 50\,000]$. Clearly, $\hat{p}$ converges to a level $A_D > 0$. The middle graph displays $t \in [0, 2000]$ showing the poor approximation by $\hat{p}$. The lower graph shows $\langle \bar{p} \rangle$ (black) for $D = 16$, $D = 12$ and $D = 8$ (top down) as well as their regression functions $\hat{p}$ (grey) on a logarithmic scale and with respect to the new asymptotic levels A_D. If $\bar{\rho}_{\text{num}}$ was constant the $\langle \bar{p} \rangle$ should be approximately parallel.

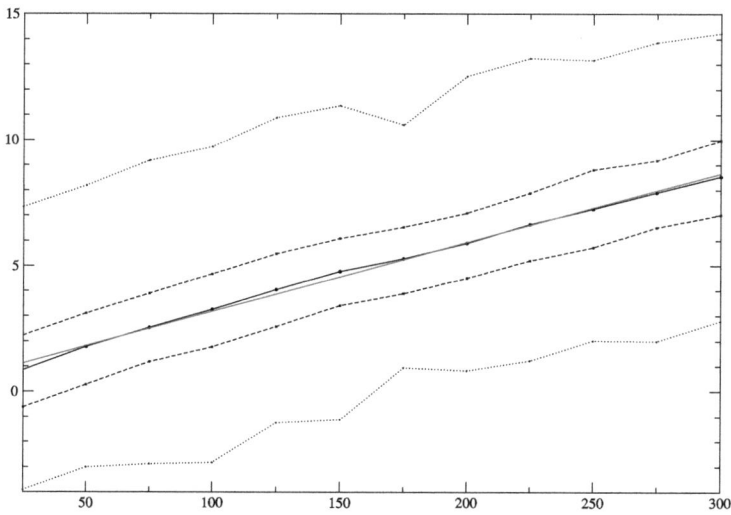

Figure 4.4: Mean, quartiles and extremal values of 500 samples of the permanent impact in dependence on the purchased volume $V \in \{25, 50, \ldots, 275, 300\}$. For every volume V, we recorded the best ask price before the trade and the averaged best ask price 500 000 steps after the trade. Here, the averaged best ask price is the mean of the best ask price sampled all 100 steps over a time interval of 100 000 steps. The linear regression of the mean is displayed in grey.

new zero line. In this case, $\bar{\rho}_{\text{num}}$ would be given by

$$\bar{\rho}_{\text{num}}(D, t) = \frac{\ln D - \ln(\langle \bar{p} \rangle_t - A_D)}{t}, \qquad (4.4.7)$$

which is fine for large t but grows to infinity as t goes to zero. To avoid this problem, we define

$$\bar{\rho}_{\text{num}}(D, t) := \frac{\ln D - \ln(\langle \bar{p} \rangle_t - (1 - e^{-t})A_D)}{t}. \qquad (4.4.8)$$

Let us point out that there is no special reason to choose $1 - \exp(-t)$.

$\langle \bar{p} \rangle$ is poorly approximated by an exponential function

Since $\langle \bar{p} \rangle$ should decay exponentially fast, $\bar{\rho}_{\text{num}}$ should be a constant. However, the existence of a permanent impact and the consequential definition of $\bar{\rho}_{\text{num}}$ in (4.4.8) makes the validity of this assumption unlikely here. Even without the permanent impact, the description of $\langle \bar{p} \rangle$ by an exponential function is poor as

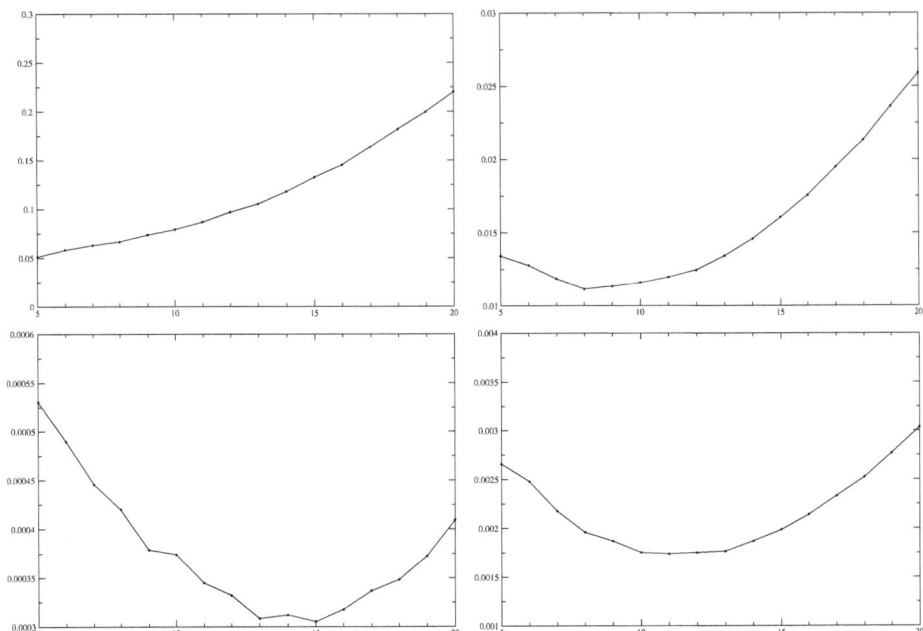

Figure 4.5: Starting in the upper left corner and proceeding clockwisely, we show the graphs of $\rho_{\text{num}}(D,\tau)$ in dependence on D for $\tau = 7, 70, 700, 7000$. Observe that the x-axis only begins in 5 due to the fact that small price impacts cannot be distinguished from the noise contained in the signal.

the upper graph of Figure 4.3 shows. As a result, $\bar{\rho}_{\text{num}}$ is time-dependent. A time-dependent resilience speed seems to be incompatible with Theorem 4.2 at first, but a closer look at the theorem's statement reveals ρ is only needed to determine the order book state *before* the next trade, given the state *after* the current trade. The time between two succeeding trades is given by τ. Thus, we focus on $\bar{\rho}_{\text{num}}(\cdot, \tau)$ and use the notation

$$\bar{\rho}_{\text{num}}(D) := \bar{\rho}_{\text{num}}(D, \tau) \qquad (4.4.9)$$

assuming that τ, which is given by the input parameters N and T, is fixed. Figure 4.5 shows the function $\bar{\rho}_{\text{num}}(\cdot, \tau)$ for several values of τ.

4.4. GAFS - A GENERALISATION OF THE AFS MODEL

$\bar{\rho}_{num}$ is a function of D

In contrast to the time dependence, the dependence on the order's price impact requires a generalisation of Theorem 4.1. Now, the resilience speed $\bar{\rho} : \mathbb{R} \to (0, \infty)$ is a continuously differentiable function of D^A. In particular, the formulas (4.2.3) and (4.2.14), which describe the price recovery in the AFS model, change to

$$D_t^A := e^{-\bar{\rho}(D_{t_n+}^A)(t-t_n)} D_{t_n+}^A \quad \text{for } t \in (t_n, t_{n+1}], \tag{4.4.10}$$

$$D_t^B := e^{-\bar{\rho}(D_{t_n+}^B)(t-t_n)} D_{t_n+}^B \quad \text{for } t \in (t_n, t_{n+1}]. \tag{4.4.11}$$

We denote this modified model as *Version 2* of the *generalised AFS model*.

For the following theorem concerning the optimal trading strategy for the GAFS model, we need two technical assumptions:

$$\text{The range of } \bar{\rho} \text{ is a subset of } [k, K], \ 0 < k < K < \infty, \text{ and} \tag{4.4.12}$$

$$1 - \tau \bar{\rho}'(x)x > 0 \text{ for all } x \in \mathbb{R}. \tag{4.4.13}$$

The first assumption bounds the resilience speed, the second assumption ensures that a larger impact cannot overtake a smaller one in the recovery phase as we will see in Lemma 4.5.

Theorem 4.3 (Optimal strategy for the generalised AFS model, Version 2). *Suppose that $\bar{\rho}$ fulfils (4.4.12) and (4.4.13), and that f satisfies*

$$\lim_{|x| \to \infty} x^2 \inf_{y \in [e^{\tau \bar{\rho}(x)} x, x]} f(y) = \infty. \tag{4.4.14}$$

Furthermore, let the function

$$h_2(x) := x \frac{f(x) - e^{-2\tau \bar{\rho}(x)} f(e^{-\tau \bar{\rho}(x)} x)(1 - \tau \bar{\rho}'(x)x)}{f(x) - e^{-\tau \bar{\rho}(x)} f(e^{-\tau \bar{\rho}(x)} x)(1 - \tau \bar{\rho}'(x)x)} \tag{4.4.15}$$

be one-to-one. Then there exists a unique optimal strategy $\xi^{(2)} = (\xi_0^{(2)}, \ldots, \xi_N^{(2)}) \in \hat{\Xi}$. The initial market order $\xi_0^{(2)}$ is the unique solution of the equation

$$F^{-1}\left(X_0 - N\left[\xi_0^{(2)} - F\left(e^{-\tau \bar{\rho}(F^{-1}(\xi_0^{(2)}))} F^{-1}(\xi_0^{(2)})\right)\right]\right) = h_2(F^{-1}(\xi_0^{(2)})), \tag{4.4.16}$$

the intermediate orders are given by

$$\xi_1^{(2)} = \cdots = \xi_{N-1}^{(2)} = \xi_0^{(2)} - F\left(e^{-\tau\bar{\rho}(F^{-1}(\xi_0^{(2)}))}F^{-1}(\xi_0^{(2)})\right), \quad (4.4.17)$$

and the final order is determined by

$$\xi_N^{(2)} = X_0 - \sum_{n=0}^{N} \xi_n^{(2)}. \quad (4.4.18)$$

In particular, the optimal strategy is deterministic. Moreover, it consists only of nontrivial buy orders, that is $\xi_n^{(2)} > 0$ for all n.

Proof. See Section 4.7.2. □

Observe that the intermediate orders of the optimal strategy, defined in (4.4.17), have the same size. Furthermore, they suggest to purchase exactly that volume that has recovered since the last trade. The GAFS model has inherited this feature from the AFS model. Yet, this observation means that also the $D_{t_n+}^A$ are equal to each other for all $n \in \{0, \ldots, N-1\}$, and thus $\bar{\rho}$ is only evaluated for one value. In other words, although $\bar{\rho}$ is a function, the optimal strategy *uses* only one value. Of course, if $\bar{\rho} \equiv \rho$ for some constant ρ in the GAFS model both models, the GAFS and the AFS, coincide. This is the main advantage of the GAFS theorem: It determines the *right* resilience speed from $\bar{\rho}$; a manual calibration, as in the AFS model, is not needed anymore.

4.4.3 From determining ρ to the GAFS model, Version 1

The procedure to determine ρ for Version 1 and the occurring problems are similar as in Section 4.4.2. Hence, we do not elaborate on the details again, but describe how we recorded the data and then focus on the GAFS model for Version 1.

Version 1 of the AFS model considers the impact onto the volume. Thus, we fixed a volume impact $E \in \{10, 20, 30, \ldots, 200\}$, and run 2500 simulations for each value. Having executed an order of size E at once, we recorded how the consumed volume recovered afterwards. In particular, the large execution took place at time $\bar{t} := 1\,000\,000$; we set

$$\bar{v} := \sum_{i=1}^{2000} \mathbb{1}_{\{p_i(\bar{t}+1)=p^a(\bar{t}+1)\}}, \qquad (4.4.19)$$

which is the volume at the best ask price after the order execution and results from the discrete nature of the model. Then, we recorded $v(t)$ defined by

$$v(t) := E - \left[\left(\sum_{i=1}^{2000} \mathbb{1}_{\{(n_i(\bar{t}+1+t)=1) \wedge (p_i(\bar{t}+1+t) \leq p^a(\bar{t}+1))\}}\right) - \bar{v}\right] \qquad (4.4.20)$$

for $t \in \{0, 2, 4, \ldots, 99\,998\}$. In comparison to our procedure for Version 2, we chose a longer time interval to look at, because the volume recovers slower than the best price. The process v is the discrete counterpart of the AFS model's process E^A.

We consider the averaged volume impact $\langle v \rangle$, pointwisely defined by

$$\langle v \rangle_t := \frac{1}{2500} \sum_{i=0}^{2500} v^i(t) \qquad (4.4.21)$$

for $t \in \{0, 2, 4, \ldots, 99\,998\}$; v^i denoting the ith sample. Let $\hat{v} : [0, \infty) \to \mathbb{R}$ be $\langle v \rangle$'s regression function of the form

$$\hat{v}_t = A + Be^{-\hat{\rho}t}, \qquad (4.4.22)$$

and let $\bar{\rho}_{\text{num}}$ denote that value for ρ that we extract from $\langle v \rangle$. Figure 4.6 shows the typical behaviour of v.

As in Version 2, we have three main problems:

1. A positive asymptote A_E,

2. a poor approximation by an exponential function, and

3. a volume-dependent $\bar{\rho}_{\text{num}}$.

The problems (1) and (2) can be treated as in the section before. In particular, we set

$$\bar{\rho}_{\text{num}}(E, t) := \frac{\ln E - \ln(\langle v \rangle_t - (1 - e^{-t})A_E)}{t} \qquad (4.4.23)$$

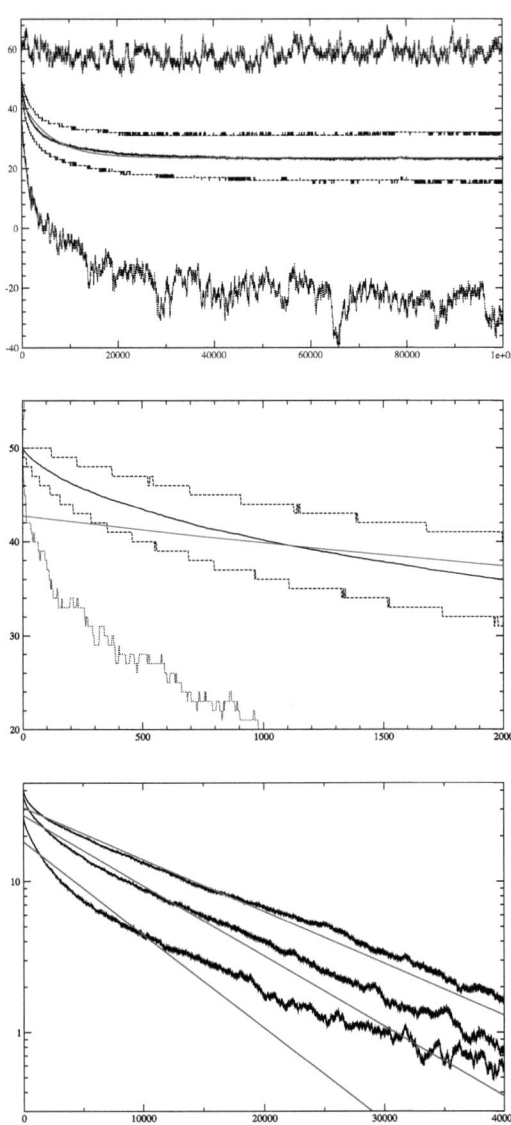

Figure 4.6: The two upper graphs show quartiles and extremal values of 2500 samples of v for $D = 50$, and the corresponding $\langle v \rangle$ and $\hat{v}$ (grey). The upper graph illustrates the long-term behaviour on the domain $t \in [0, 100\,000]$; the middle graph displays $t \in [0, 2000]$. The lower graph shows $\langle v \rangle$ (black) for $D = 150$, $D = 100$ and $D = 50$ (top down) as well as their regression functions $\hat{v}$ (grey) on a logarithmic scale and with respect to the new asymptotic levels A_E.

4.4. GAFS - A GENERALISATION OF THE AFS MODEL

and point out that Theorem 4.1 is only interested in

$$\bar{\rho}_{\text{num}}(E) := \bar{\rho}_{\text{num}}(E, \tau). \qquad (4.4.24)$$

For problem (3), we generalise the AFS model, Version 1, to the GAFS model, Version 1. Now, the resilience speed $\bar{\rho} : [0, \infty) \to (0, \infty)$ is a *twice* differentiable function of E^A. In particular, the formulas (4.2.10) and (4.2.14) from the AFS model become

$$E_t^A := e^{-\bar{\rho}(E_{t_n+}^A)(t-t_n)} E_{t_n+}^A, \ t \in (t_n, t_{n+1}], \qquad (4.4.25)$$

$$E_t^B := e^{-\bar{\rho}(E_{t_n+}^B)(t-t_n)} E_{t_n+}^B, \ t \in (t_n, t_{n+1}]. \qquad (4.4.26)$$

in the GAFS model. Then, the following theorem determines the optimal trading strategy in the set of all admissible strategies $\hat{\Xi}$:

Theorem 4.4 (Optimal strategy for the generalised AFS model, Version 1). *Suppose that $\bar{\rho}$ fulfils the assumptions (4.4.12) and (4.4.13), and additionally*

$$e^{-\bar{\rho}(x)\tau}\left(1 - \tau\bar{\rho}'(x)x\right) < 1 \ \text{for all } x \in \mathbb{R}. \qquad (4.4.27)$$

Furthermore, let the function

$$h_1(x) := \frac{F^{-1}(x) - e^{-\bar{\rho}(x)\tau}\left(1 - \tau\bar{\rho}'(x)x\right) F^{-1}(e^{-\bar{\rho}(x)\tau}x)}{1 - e^{-\bar{\rho}(x)\tau}\left(1 - \tau\bar{\rho}'(x)x\right)} \qquad (4.4.28)$$

be one-to-one. Then there exists a unique optimal strategy $\xi^{(1)} = (\xi_0^{(1)}, \ldots, \xi_N^{(1)}) \in \hat{\Xi}$. The initial market order $\xi_0^{(1)}$ is the unique solution of the equation

$$F^{-1}\left(X_0 - N\xi_0^{(1)}(1 - e^{-\bar{\rho}(\xi_0^{(1)})\tau})\right) = h_1(\xi_0^{(1)}), \qquad (4.4.29)$$

the intermediate orders are given by

$$\xi_1^{(1)} = \cdots = \xi_{N-1}^{(1)} = \xi_0^{(1)}\left(1 - e^{-\bar{\rho}(\xi_0^{(1)})\tau}\right), \qquad (4.4.30)$$

and the final order is determined by

$$\xi_N^{(1)} = X_0 - \sum_{n=0}^{N} \xi_n^{(1)}. \qquad (4.4.31)$$

N	T	$\xi_0^{(2)}$	$\xi_1^{(2)}$	$\xi_N^{(2)}$	Predicted	Sampled	Samp/Pred
40	400	8.95	4.74	6.38	701.47	1867.74	266%
40	4000	6.13	4.81	6.15	500.24	1573.50	315%
40	40 000	5.16	4.86	5.40	392.42	1076.89	274%
50	400	8.29	3.80	5.29	691.94	1853.37	268%
50	4000	5.20	3.88	4.94	462.51	1535.96	332%
50	40 000	4.26	3.90	4.82	349.26	1014.42	290%
80	400	7.55	2.37	5.63	691.65	1832.69	265%
80	4000	3.73	2.44	3.37	387.98	1464.03	377%
80	40 000	2.65	2.46	2.91	231.67	914.17	395%

Table 4.1: The optimal strategies according to the GAFS model, Version 2, for $X = 200$ and several values for N and T.

In particular, the optimal strategy is deterministic. Moreover, it consists only of nontrivial buy orders, that is $\xi_n > 0$ for all n.

Proof. See Section 4.7.1. □

As in Version 2 of the GAFS model, $\bar{\rho}$ is only evaluated in one value, and if $\bar{\rho} \equiv \rho$ the best strategies of the GAFS and the AFS models coincide.

4.5 The (G)AFS optimal strategies applied to the Opinion Game

Let us turn to the numerical results of this chapter. We use the parameter values determined in the last section to calculate the GAFS optimal strategies and apply them in the Opinion Game. We show first that the resulting costs show an *expected* behaviour on a general level, and that the AFS model with a suboptimal value for ρ suggests a strategy that produces significantly higher costs than the corresponding GAFS strategy. Afterwards, we compare the costs sampled in the Opinion Game to the costs predicted by the GAFS model and find significant differences. Again, we first treat Version 2 in detail in Section 4.5.1 and, afterwards, Version 1 in Section 4.5.2.

4.5.1 Results for Version 2

Here, we refer to the values for f and $\bar{\rho}$, $\bar{\rho}_{\text{num}}$, as determined in the Sections 4.4.1 and 4.4.2.

Strategy	$\xi_0^{(2)}$	$\xi_1^{(2)}$	$\xi_N^{(2)}$	Predicted	Sampled
GAFS	3.73	2.44	3.37	387.98	1464.03
AFS	21.02	0.97	102.26	979.97	1584.12

Table 4.2: The optimal strategies and their costs for the AFS model with $\rho = \hat{\rho}$ and the GAFS model with $\bar{\rho}$ from section 4.4.2. $(X, T, N) = (200, 4000, 80)$.

Table 4.1 shows the GAFS optimal strategies and their costs for different values of T and N. We consider two kinds of costs. The *predicted costs* are the impact costs that are theoretically predicted by the (G)AFS model. Here, we assume that the market behaves as described in Section 4.2. The *sampled costs* are the average of 500 samples with the given strategy in the Opinion Game. Observe first that the predicted and sampled costs decrease if the trading time or the number of trading opportunities increase. Of course, this is no special feature of the (G)AFS strategies; every fixed strategy benefits from a larger τ, which is implied by a greater T, and additional trading opportunities can be used, but do not have to be used. Thus, every reasonable strategy can only perform better with larger T or N. Nevertheless, the costs of the GAFS strategies show a *reasonable* behaviour.

Furthermore, the GAFS strategies perform better than the AFS strategies: Recall that the AFS model with the right value for ρ results in the same optimal strategy as the GAFS model. Moreover, the (G)AFS model assumes an exponential decay of the price impact (see (4.4.4)). We have taken this assumption into account by introducing $\langle \bar{\rho} \rangle$'s regression function $\hat{p}$ in (4.4.6), which was of the form

$$\hat{p}_t := A + Be^{-\hat{\rho}t}. \tag{4.5.1}$$

Table 4.2 shows the optimal strategies and their costs for $(X, T, N) = (200, 4000, 80)$ with respect to the AFS model with $\rho = \hat{\rho}$ and the GAFS model with $\bar{\rho} = \bar{\rho}_{\text{num}}$. The example shows that a naive guess of a good ρ can lead to much higher costs: The AFS costs amount 253% of the GAFS costs in prediction, and still 108% in the samples.

The last two paragraphs have shown that the GAFS strategies are reasonable and superior to the AFS strategies. However, returning to Table 4.1, we see

N	T	$\xi_0^{(2)}$	$\xi_1^{(2)}$	$\xi_N^{(2)}$	Predicted	Sampled	Samp/Pred
40	400	16.95	4.41	11.17	1002.02	1846.48	184%
40	4000	12.76	4.51	11.37	872.63	1565.66	179%
40	40000	7.17	4.75	7.72	591.42	1064.72	180%
50	400	14.96	3.56	10.71	959.38	1837.80	192%
50	4000	11.61	3.67	8.66	845.09	1532.69	181%
50	40000	6.29	3.85	5.05	562.05	1012.16	180%
80	400	11.48	2.30	7.12	863.04	1869.49	217%
80	4000	9.62	2.33	6.16	794.95	1479.48	186%
80	40000	4.16	2.30	13.89	469.40	940.49	200%

Table 4.3: The optimal strategies according to the GAFS model, Version 1, for $X = 200$ and several values for N and T.

that the predicted and the sampled costs for the individual parameter sets differ strongly from each other. The last column shows both kinds of costs in relation to each other. Obviously, the sampled costs are multiple times higher. This observation is a strong evidence that the assumptions of the (G)AFS model are insufficient to capture the whole complexity of the order book dynamics in the Opinion Game. It is doubtful if the (G)AFS model really suggests optimal trading strategies for this artificial market environment. With regard to the Opinion Game features concerning the order book behaviour that we have discussed in Section 4.3, it is unlikely that the (G)AFS strategies minimise the costs in real world markets.

4.5.2 Results for Version 1

Recall that Version 1 of the GAFS model assumes an exponentially fast recovery of the volume. Table 4.3 shows the optimal strategies for several values of N and T. Again, the predicted and sampled costs decrease with an increasing number of trading opportunities, N, or a longer trading horizon, T. Nevertheless, also Version 1 shows large differences of the sampled costs compared to the predicted once. Even if not as large as for Version 2, the sampled costs are approximately twice as high. The slightly better performance can be seen as a hint that the assumptions of Version 1 are closer to the order book dynamics in the Opinion Game; yet, there is still a big gap between prediction and samples showing that, also in Version 1, important features of the market dynamics are missing in the (G)AFS model.

4.6 Missing features of the (G)AFS model

> *Everything I said was true,*
> *but I couldn't prove it.*
> Maximo Park - Postcard of a Painting

The last section has shown that the theoretically predicted costs differ strongly from the sampled costs. Obviously, the dynamics of the Opinion Game are not sufficiently well captured by the (G)AFS model. In this section, we discuss some of the missing features. The discussion is not exhaustive, but serves as collage of various ideas how to improve the (G)AFS model for the Opinion Game market. Of course, the Opinion Game is also just an artificial trading environment; an improvement of the (G)AFS model with respect to this market does not imply that the improved model would also perform better on real financial markets. Nevertheless, we will see that the suggested features are by no means exotic, but are also known for real markets or seem to be reasonable at least.

The first reason for the differences between predicted and sampled costs is the lack of permanent impact in the (G)AFS model. The assumed linearity of the permanent impact (see page 95) sometimes leads to the argumentation that it does not influence the best strategy as the nth traded share always suffers from the permanent impact of the $(n-1)$ shares before, no matter what strategy is chosen. A closer look at the method of measurement shows that this assumption is wrong. The linearity rather refers to the impact of large orders on the next traded block. In this sense, we may speak of a *blockwise* linearity, and it is reasonable to ask if the optimal strategy minimising the average cost (4.2.16) is at least close to the optimal strategy minimising the average cost including blockwise linear permanent impact,

$$\mathbb{E}\left(\sum_{n=0}^{N}\left[\pi_{t_n}(\xi_n) + \lambda \sum_{m=0}^{n-1}\xi_m\right]\right) \qquad (4.6.1)$$

with λ being the slope of the permanent impact function. For the Opinion Game, the introduction of a blockwise linear impact into the (G)AFS market model leads to a better match of predicted and sampled costs. Figure 4.7 supports this

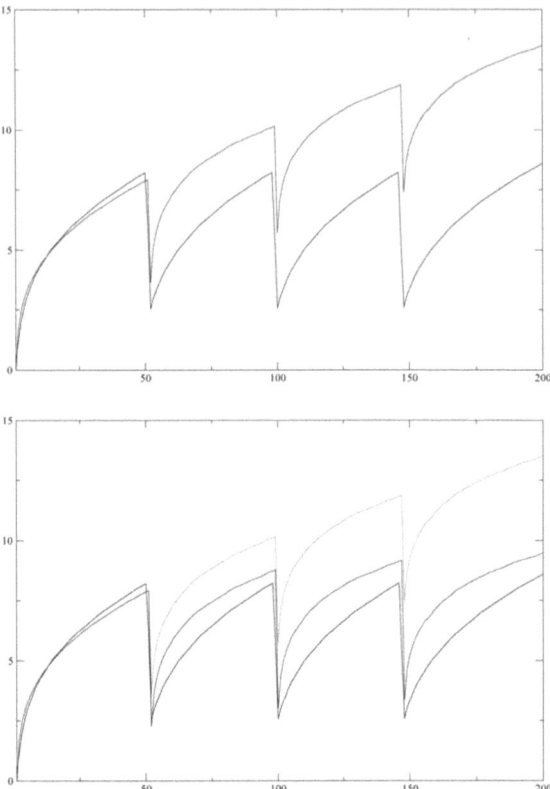

Figure 4.7: The impact costs per unit for $X = 200$, $N = 3$, $T = 2100$. The black, regular graph shows the costs as predicted by the GAFS model, Version 2. In the upper figure, the dark grey graph shows the sampled costs; in the lower figure, it is coloured light grey. There, the dark grey graph displays the sampled costs without blockwise permanent impact.

statement. It shows the predicted and sampled costs for every single bought share of a large order. One can clearly see that the prediction and the sample match well until the first recovery period is over. Afterwards, the sampled costs suffer from an approximately linear increase, which can be interpreted as the permanent impact. Assuming that this impact is, indeed, a linear function of the order volume, we can use our knowledge of the function's slope from Section 4.4.2 and the order book's shape from Section 4.4.1 to remove it. The resulting graph fits much better the predicted shape.

4.6. MISSING FEATURES OF THE (G)AFS MODEL

Let us take another look at the upper illustration of Figure 4.7. We first consider the predicted costs, displayed in black. Their graph has the shape of three perfectly identical teeth; only the last one is slightly larger. This picture is not surprising. Every bought share is more expensive than the previous one, the costs increase. Then, there is the first trading break, the order book recovers, and the first share that is bought after the break is cheaper. As the (G)AFS model assumes a static order book shape, the price increase is identical to the previous one. Only the last order's volume is larger such that the costs increase more than before.

Let us now *toothwisely* compare the black graph to the dark grey graph, which show the sampled, *permanent impact free* costs. While the first teeth are very similar to each other, the following teeth of the sampled costs are more concave than their predicted counterparts. This means the costs per share increase faster in the beginning of an order execution, inducing less supply in vicinity of the best ask price. On the other hand, the sampled costs increase slower in the end of an execution. A possible economic interpretation is that, after the first order, traders that have observed the rise in the price hope for a repeating. They therefore place orders around the highest price that they have observed before, providing additional supply and, consequently, more resistance against a further rise of the price. Nevertheless, the main observation is the dynamic nature of the order book shape.

Observe that Figure 4.7 shows the costs of the (G)AFS model's Version 2, which assumes an exponential decay of the price impact. Comparing the minima of the black and the red graph in the right illustration demonstrates that our choice of $\bar{\rho}$ is quite good as the minima, the best ask prices after the recovery phases, approximately match. For a better understanding of how the supply at the best price looks like, we refer to Figure 4.8, which shows a dynamic trading strategy. While the size of the first order is given by the optimal (G)AFS strategy, the following orders have just the volume that has recovered in the trading break. The remaining volume is purchased in the last step. According to the (G)AFS

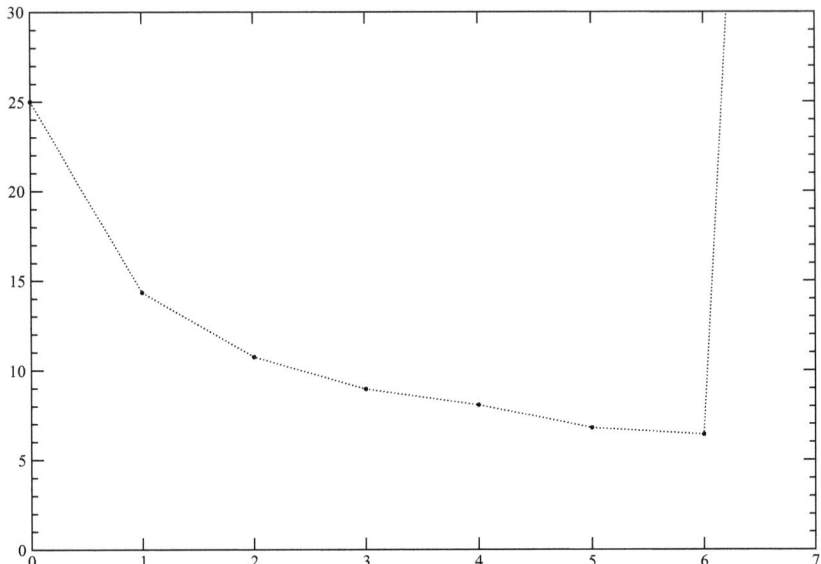

Figure 4.8: Averaged behaviour of a dynamical strategy in the Opinion Game: Only the recovered volume is purchased.

model, the middle orders should have the same size, but the figure demonstrates that this is not the case: The recovering volume is decreasing. Consequently, the best price may recover as predicted, but the supply there and in vicinity is low. We can interpret this observation by a learning effect, which complements the interpretation from the last paragraph: The more often the agents observe that the price periodically rises to a certain level, the less they are willing to place orders below it.

Let us summarise the observed features of the Opinion Game that are not captured by the (G)AFS model:

- Blockwise linear permanent impact;
- less volume at the best quote after the recovery phases,
- but more liquidity at higher prices;
- deceleration of the volume's recovery speed with every additional executed trading block.

We have shown in (4.6.1) how the first item can be implemented into the (G)AFS model. The implementation of the other features is more difficult, since we do not have a detailed, quantitative description of the qualitative behaviour. Besides the modelling, every additional feature in the (G)AFS model also increases the difficulties in proving optimal trading strategies; already the implementation of a permanent impact let the proofs of the Theorems 4.3 and 4.4 fail. More research is necessary here.

4.7 Proofs of the Theorems 4.3 and 4.4

The structure of the proofs remains the same as in the proofs of the corresponding AFS theorems (see Appendices A to C in Alfonsi et al. (2010)). Nevertheless, we need to justify the constraints on $\bar{\rho}$; furthermore, the computations become more complicated by our generalisation. For simplicity, we assume $t_0 = 0$ in this section.

We start with the introduction of slightly changed dynamics for the GAFS model and the reduction of the admissible strategies to deterministic ones. For any admissible strategy ξ, the new dynamics is defined by the processes $D := (D_t)_{t \geq 0}$ and $E := (E_t)_{t \geq 0}$. We set $D_0 = D_{t_0} := 0 =: E_{t_0} = E_0$ and

$$E_{t_n+} := E_{t_n} + \xi_n \text{ and } D_{t_n+} := F^{-1}(F(E_{t_n}) + \xi_n) \quad (4.7.1)$$

for the trading times $t_0, \ldots, t_N$. The processes' values between two successive trading times $t \in (t_n, t_{n+1})$ are given by

$$\begin{aligned} E_t &:= e^{-\bar{\rho}(E_{t_n+})(t-t_n)} E_{t_n+} \quad \text{for Version 1;} \\ D_t &:= e^{-\bar{\rho}(D_{t_n+})(t-t_n)} D_{t_n+} \quad \text{for Version 2.} \end{aligned} \quad (4.7.2)$$

Given one process, we can recover the other one by the equations (4.2.8):

$$E_t = F(D_t) \text{ and } D_t = F^{-1}(E_t). \quad (4.7.3)$$

Lemma 4.5. *Under assumption (4.4.13),*

$$E_t^B \leq E_t \leq E_t^A \text{ and } D_t^B \leq D_t \leq D_t^A \quad (4.7.4)$$

for all $t \geq 0$. In the special case that all ξ_n are nonnegative, we have $D^A = D$ and $E^A = E$.

Proof. To see that $D^A = D$ and $E^A = E$ if ξ consists of buy orders only, observe that the new dynamics match exactly the original ones for such a ξ.

For the general case, we consider $E^B_t \leq E_t$; the other inequalities follow equivalently. Observe that it is sufficient to prove

$$E^B_{t_{n+1}} \leq E_{t_{n+1}} \tag{4.7.5}$$

for

$$E^B_{t_n+} \leq E_{t_n+}, \tag{4.7.6}$$

since the functions are both exponentially decreasing on $(t_n, t_{n+1}]$, and the relative order of E^B and E cannot be reversed from t_n to t_n+. Furthermore, we can restrict ourselves to the case that $E^B_{t_n+}$ and E_{t_n+} have the same sign, as the signs cannot change in the considered time interval, $(t_n, t_{n+1}]$. We are consequently done if we can show that the inequality

$$|x+y|e^{-\bar{\rho}(x+y)\tau} > |x|e^{-\bar{\rho}(x)\tau} \tag{4.7.7}$$

holds for all $(x, y) \in \mathbb{R}^2$ with $\text{sgn}(x+y) = \text{sgn}(x)$ and $|x+y| > |x|$. Observe that we have equality in the equation above if we consider the trivial case that $y = 0$. We define a function $u_x : \mathbb{R} \to \mathbb{R}$ by

$$u_x(y) := (x+y)e^{-\bar{\rho}(x+y)\tau}. \tag{4.7.8}$$

Differentiation yields

$$u'_x(y) = e^{-\bar{\rho}(x+y)\tau}(1 - \tau \bar{\rho}'(x+y)(x+y)). \tag{4.7.9}$$

The right-hand side of this equation is positive by assumption (4.4.13), thus u_x is strictly increasing. Since $u_x(0) = xe^{-\bar{\rho}(x)\tau}$, (4.7.7) is proven. $\square$

4.7. PROOFS OF THE THEOREMS 4.3 AND 4.4

It remains to define the *simplified price of* ξ_n *under the new dynamics* by

$$\bar{\pi}_{t_n}(\xi_n) := \int_{D_{t_n}}^{D_{t_n}+} (A_{t_n}^0 + x) f(x) dx = A_{t_n}^0 \xi_n + \int_{D_{t_n}}^{D_{t_n}+} x f(x) dx. \quad (4.7.10)$$

Observe that

$$\bar{\pi}_{t_n}(\xi_n) \leq \pi_{t_n}(\xi_n) \quad (4.7.11)$$

for all admissible strategies ξ because of Lemma 4.5. In particular, if ξ consists of buy orders only, we have equality.

We show in the next two sections that the strategies given in the Theorems 4.4 and 4.3, $\xi^{(1)}$ and $\xi^{(2)}$, are the unique minimisers of the *price functional*

$$\bar{\mathscr{C}}(\xi) := \mathbb{E}\left(\sum_{n=0}^{N} \bar{\pi}_{t_n}(\xi_n)\right) \quad (4.7.12)$$

for the corresponding version of the model. As $\xi^{(1)}$ and $\xi^{(2)}$ consist of buy orders only, (4.7.11) and the remark afterwards imply that these strategies are also the minimisers of the original price functional $\mathscr{C}$.

We turn to the reduction of $\hat{\Xi}$ to deterministic strategies. Let us define the *remaining trading volume* $X = (X_t)_{t \in [0,T]}$ by

$$X_t := X_0 - \sum_{t_n < t} \xi_n. \quad (4.7.13)$$

Furthermore, we set $X_{t_{N+1}} := 0$. We can transform the price of a strategy $\xi \in \hat{\Xi}$ by

$$\sum_{n=0}^{N} \bar{\pi}_{t_n}(\xi_n) = \sum_{n=0}^{N} A_{t_n}^0 \xi_n + \sum_{n=0}^{N} \int_{D_{t_n}}^{D_{t_n}+} x f(x) dx, \quad (4.7.14)$$

and use definition (4.7.13) as well as *integration by parts* to rewrite the first term on the right-hand side:

$$\sum_{n=0}^{N} A_{t_n}^0 \xi_n = -\sum_{n=0}^{N} A_{t_n}^0 \left(X_{t_{n+1}} - X_{t_n}\right) = X_0 A_0 + \sum_{n=1}^{N} X_{t_n} \left(A_{t_n}^0 - A_{t_{n-1}}^0\right). \quad (4.7.15)$$

Since ξ is admissible, X is a bounded process and X_{t_n} is $\mathcal{F}_{t_{n-1}}$-measurable. A^0 is a martingale, thus, the expectation of (4.7.15) must be $X_0 A_0$. The second term on the right-hand side of (4.7.14) is deterministic for a given realisation $\xi(\omega)$ of a strategy ξ. We denote this term by

$$C^{(i)}(\xi): \mathbb{R}^{N+1} \to \mathbb{R}$$
$$\xi \mapsto \sum_{n=0}^{N} \int_{D_{t_n}}^{D_{t_n+}} x f(x) dx \qquad (4.7.16)$$

for Version i, $i \in \{1, 2\}$. Now, we can express $\bar{\mathscr{C}}$ by

$$\bar{\mathscr{C}}(\xi) = A_0 X_0 + \mathbb{E}(C^{(i)}(\xi)). \qquad (4.7.17)$$

We spend the next two sections to show $C^{(i)}$ has a unique minimiser in the set

$$\Xi := \left\{ x := (x_0, \ldots, x_N) \in \mathbb{R}^{N+1} : \sum_{n=0}^{N} x_n = X_0 \right\} \qquad (4.7.18)$$

and this minimiser is determined by the formula given in Theorem 4.3 or 4.4 respectively.

For the sake of convenience, we introduce some more notation:

$$\bar{a}_x := \exp(-\tau \bar{\rho}(x)) \text{ for } x \in \mathbb{R}, \qquad (4.7.19)$$

$$a_n := \begin{cases} \exp(-\tau \bar{\rho}(E_{t_n+})) & \text{in Section 4.7.1} \\ \exp(-\tau \bar{\rho}(D_{t_n+})) & \text{in Section 4.7.2} \end{cases} \text{ for } n \in \{0, \ldots, N\}. \qquad (4.7.20)$$

Because the range of $\bar{\rho}$ is assumed to be $[k, K]$, $0 < k < K < \infty$, by (4.4.12),

$$e^{-\tau K} \leq \bar{a}_x \leq e^{-\tau k} \text{ and } e^{-\tau K} \leq a_n \leq e^{-\tau k}. \qquad (4.7.21)$$

Additionally, we will need these functions:

$$\tilde{F}(x) := \int_0^x z f(z) dz \text{ and } G(x) := \tilde{F}(F^{-1}(x)). \qquad (4.7.22)$$

Observe that

$$G'(x) = \tilde{F}'(F^{-1}(x))(F^{-1})'(x) = F^{-1} f(F^{-1}(x)) \frac{1}{f(F^{-1}(x))} = F^{-1}(x), \qquad (4.7.23)$$

and thus G is twice continuously differentiable, nonnegative, convex and has a fixed point in 0.

4.7. PROOFS OF THE THEOREMS 4.3 AND 4.4

4.7.1 The optimal strategy for Version 1

In this section, we calculate the unique minimiser of $C^{(1)}$ in Ξ. For any $\xi = (x_0, \ldots, x_N) \in \Xi$, we have

$$C^{(1)}(\xi) = \sum_{n=0}^{N} \int_{D_{t_n}}^{D_{t_n+}} x f(x) dx \qquad (4.7.24)$$

$$= \sum_{n=0}^{N} \left[\tilde{F}(F^{-1}(E_{t_n+})) - \tilde{F}(F^{-1}(E_{t_n})) \right] \qquad (4.7.25)$$

$$= \sum_{n=0}^{N} \left[G(E_{t_n} + x_n) - G(E_{t_n}) \right]. \qquad (4.7.26)$$

Lemma 4.6. *The function $C^{(1)}$ has at least one local minimum in Ξ.*

Proof. The statement will follow from

$$C^{(1)}(\xi) \to \infty \text{ for } ||\xi||_\infty \to \infty, \qquad (4.7.27)$$

because $C^{(1)}$ is continuous. First, we use the properties of G to find a lower bound for $G(x) - G(cx)$ with $x \in \mathbb{R}$ and $c \in [0,1]$:

$$G(x) - G(cx) \geq G(cx) + (x - cx)G'(cx) - G(cx) \qquad (4.7.28)$$

$$= (1-c)|F^{-1}(cx)||x|. \qquad (4.7.29)$$

The inequality (4.7.28) applied to (4.7.26) leads to a lower bound for $C^{(1)}$:

$$\begin{aligned}
C^{(1)}(\xi) &= G(E_{t_N} + x_N) - G(E_{t_0}) + \sum_{n=0}^{N-1} \left[G(E_{t_n} + x_n) - G(E_{t_{n+1}}) \right] \quad (4.7.30) \\
&= G\left(\left(\Pi_{n=0}^{N} a_n\right) x_0 + \cdots + a_{N-1} x_{N-1} + x_N \right) - G(0) \\
&\quad + \sum_{n=0}^{N-1} \left[G\left(\left(\Pi_{m=0}^{n-1} a_m\right) x_0 + \cdots + a_{n_1} x_{n-1} + x_n \right) \right. \\
&\quad \left. - G\left(a_n \left[\left(\Pi_{m=0}^{n-1} a_m\right) x_0 + \cdots + a_{n_1} x_{n-1} + x_n \right] \right) \right] \quad (4.7.31) \\
&\overset{(4.7.26)}{\geq} G\left(\left(\Pi_{n=0}^{N} a_n\right) x_0 + \cdots + a_{N-1} x_{N-1} + x_N \right) - G(0) \\
&\quad + \sum_{n=0}^{N-1} \left[(1-a_n) \left| F^{-1}\left(a_n \left[\left(\Pi_{m=0}^{n-1} a_m\right) x_0 + \cdots + a_{n_1} x_{n-1} + x_n \right] \right) \right| \right. \\
&\quad \left. \left| \left[\left(\Pi_{m=0}^{n-1} a_m\right) x_0 + \cdots + a_{n_1} x_{n-1} + x_n \right] \right| \right]. \quad (4.7.32)
\end{aligned}$$

We define a linear mapping $T : \mathbb{R}^{N+1} \to \mathbb{R}^{N+1}$ by

$$T(\xi) := \left(x_0, a_0 x_0 + x_1, \ldots, \left[\Pi_{n=0}^{N-1} a_n\right] x_0 + \cdots + a_{N-1} x_{N-1} + x_N\right), \quad (4.7.33)$$

and the smallest a_n by

$$a := \min\{a_n : n \in \{0, \ldots, N\}\}. \quad (4.7.34)$$

Observe that

$$\|T(\xi)\|_\infty \geq \|(x_0, a x_0 + x_1, \ldots, a^n x_0 + \cdots + a x_{N-1} + x_N)\|_\infty \to \infty \quad (4.7.35)$$

for $\|\xi\|_\infty \to \infty$, as well as $G(x) \to \infty$ and $|F^{-1}(ax)||x| \to \infty$ for $|x| \to \infty$. The last statement follows, because F is unbounded. Finally, we define

$$H(x) := \min\left(G(x), |F^{-1}(ax)||x|\right). \quad (4.7.36)$$

Also $H(x) \to \infty$ for $|x| \to \infty$, and consequently,

$$C^{(1)}(\xi) \geq H(\|T(\xi)\|_\infty) - G(0) \to \infty. \quad (4.7.37)$$

$\square$

One has to determine $\xi_0^{(1)}$ by solving

$$F^{-1}\left(X_0 - N\xi_0^{(1)}(1 - a_0)\right) = h_1(\xi_0^{(1)}) \quad (4.7.38)$$

in Theorem (4.4). We define the function

$$\hat{h}_1(x) := h_1(x) - F^{-1}\left(X_0 - N(1 - \bar{a}_x)x\right) \quad (4.7.39)$$

for which $\xi_0^{(1)}$ is a zero.

Lemma 4.7. *Given that the assumptions of Theorem 4.4 hold, function $\hat{h}_1$ has at most one zero, which is positive if it exists.*

Proof. For the existence of at most one zero, it is sufficient to show that $\hat{h}_1$ is strictly increasing. The function h_1 has a fixed point in 0, is positive for positive arguments and continuous as well as bijective, thus it must be strictly increasing or, equivalently, its slope must be strictly positive. Consequently, the slope of $\hat{h}_1$ is also positive, because

$$\hat{h}_1'(x) = h_1'(x) - \frac{d}{dx}\left[F^{-1}\left(X_0 - Nx(1-\bar{a}_x)\right)\right] \quad (4.7.40)$$

$$= h_1'(x) + N\frac{1-\bar{a}_x\left(1-\tau\bar{\rho}'(x)x\right)}{f(F^{-1}(X_0 - Nx(1-\bar{a}_x)))}, \quad (4.7.41)$$

and the numerator of the second term is positive by assumption (4.4.27). The positivity of the zero (if existing) follows simply from

$$\hat{h}_1(0) = -F^{-1}(X_0) < 0. \quad (4.7.42)$$

$\square$

Now, we are prepared to prove Theorem 4.4.

Lemma 4.8. *Strategy $\xi^{(1)}$ is the unique minimiser of function $C^{(1)}$ and all components of $\xi^{(1)}$ are positive.*

Proof. We showed in Lemma 4.6 that there is an optimal strategy $\xi^* = (x_0^*, \ldots, x_N^*) \in \Xi$. Thus, there must be a Lagrange multiplier $\nu \in \mathbb{R}$ such that

$$\frac{\partial}{\partial x_n^*}C^{(1)}(\xi^*) = \nu \quad \text{for } n \in \{0, \ldots, N\}. \quad (4.7.43)$$

Using representation (4.7.26) of $C^{(1)}$, one gets

$$\frac{\partial}{\partial x_n}C^{(1)}(x) = F^{-1}(E_{t_n+}) + a_n\left(1 - \bar{\rho}'(E_{t_n+})E_{t_n+}\right)\left[\frac{\partial}{\partial x_{n+1}}C^{(1)}(x) - F^{-1}(E_{t_{n+1}})\right] \quad (4.7.44)$$

for $n \in \{0, \ldots, N-1\}$. In combination with the Langrange multiplier, the recursive formula yields

$$F^{-1}(E_{t_n+}) + a_n\left(1 - \bar{\rho}'(E_{t_n+})E_{t_n+}\right)\left[\nu - F^{-1}(E_{t_{n+1}})\right] = \nu \quad (4.7.45)$$

$$\Leftrightarrow \nu = \frac{F^{-1}(E_{t_n+}) - a_n\left(1 - \bar{\rho}'(E_{t_n+})E_{t_n+}\right)F^{-1}(a_n E_{t_n+})}{1 - a_n\left(1 - \bar{\rho}'(E_{t_n+})E_{t_n+}\right)} = h_1(E_{t_n+}) \quad (4.7.46)$$

for $n \in \{0, \ldots, N-1\}$. The function h_1 is bijective by assumption, and thus

$$x_0^* = h_1^{-1}(\nu) \tag{4.7.47}$$
$$x_n^* = (1-a_0)x_0^* \text{ for } n \in \{1, \ldots, N-1\} \tag{4.7.48}$$
$$x_N^* = X_0 - x_0^* - (N-1)x_0^*(1-a_0). \tag{4.7.49}$$

Therefore, the optimal strategy ξ^* is completely defined if we can determine x_0^*. By (4.7.26),

$$\begin{aligned} C^{(1)}(x^*) &= G(x_0^*) - G(0) + (N-1)\left[G(a_0 x_0^* + (1-a_0)x_0^*) - G(a_0 x_0^*)\right] \\ &\quad + G(a_0 x_0^* + X_0 - x_0^* - (N-1)(1-a_0)x_0^*) - G(a_0 x_0^*) \quad (4.7.50) \\ &= N\left[G(x_0^*) - G(a_0 x_0^*)\right] + G(X_0 - N(1-a_0)x_0^*) - G(0) \quad (4.7.51) \\ &=: C_0^{(1)}(x_0^*). \tag{4.7.52} \end{aligned}$$

We know that $C_0^{(1)}$ has a minimum because of Lemma 4.6; we can find it by differentiation:

$$\begin{aligned} \frac{d}{dx} C_0^{(1)}(x) &= N\big[F^{-1}(x) - \bar{a}_x(1-\tau\bar{\rho}'(x)x)F^{-1}(\bar{a}_x x) \\ &\quad - \left[1 - \bar{a}_x(1-\tau\bar{\rho}'(x)x)\right]F^{-1}(X_0 - N(1-\bar{a}_x)x)\big] \quad (4.7.53) \\ &= N\left[1 - \bar{a}_x(1-\tau\bar{\rho}'(x)x)\right]\hat{h}_1(x). \tag{4.7.54} \end{aligned}$$

Assumption (4.4.27) and Lemma 4.7 tell us $C^{(1)}$ has exactly one minimum, and this minimum is positive. We have established the uniqueness and representation of the optimal strategy.

It remains to show that all components of x^* are positive. We already know that $x_0^* > 0$ by Lemma 4.7. The positivity of x_n^* follows from (4.7.48) for all $n \in \{1, \ldots, N-1\}$. For the last order, x_N^*, observe that (4.7.53) vanishes in x_0^*:

$$\begin{aligned} 0 &= F^{-1}(x_0^*) - a_0(1-\tau\bar{\rho}'(x_0^*)x_0^*)F^{-1}(a_0 x_0^*) \\ &\quad - \left[1 - a_0(1-\tau\bar{\rho}'(x_0^*)x_0^*)\right]F^{-1}(\underbrace{X_0 - N(1-a_0)x_0^*}_{=x_N^* + a_0 x_0^*}). \tag{4.7.55} \end{aligned}$$

Furthermore, F^{-1} is strictly increasing, and thus

$$0 > \left[1 - a_0(1 - \tau\bar{\rho}'(x_0^*)x_0^*)\right]\left[F^{-1}(a_0 x_0^*) - F^{-1}(a_0 x_0^* + x_N^*)\right], \quad (4.7.56)$$

which, indeed, implies the positivity of x_N^*. □

4.7.2 The optimal strategy for Version 2

In this section, we determine the unique minimiser of $C^{(2)}$ in Ξ. For $\xi = (x_0, \ldots, x_N) \in \Xi$, we have

$$C^{(2)}(\xi) = \sum_{n=0}^{N} \int_{D_{t_n}}^{D_{t_n+}} x f(x) dx \quad (4.7.57)$$

$$= \sum_{n=0}^{N} \left(G(x_n + F(D_{t_n})) - \tilde{F}(D_{t_n}) \right) \quad (4.7.58)$$

Lemma 4.9. *The function $C^{(2)}$ has a local minimum in Ξ.*

Proof. Again, it suffices to show

$$C^{(2)}(\xi) \to \infty \text{ for } ||\xi||_\infty \to \infty. \quad (4.7.59)$$

We rearrange (4.7.58) and get

$$C^{(2)}(\xi) \quad (4.7.60)$$

$$= \sum_{n=0}^{N} \left(\tilde{F}(F^{-1}(x_n + F(D_{t_n}))) - \tilde{F}(D_{t_n}) \right) \quad (4.7.61)$$

$$= \tilde{F}(a_N F^{-1}(x_N + F(D_{t_N})))$$

$$+ \sum_{n=0}^{N} \left(\tilde{F}(F^{-1}(x_n + F(D_{t_n}))) - \tilde{F}(a_n F^{-1}(x_n + F(D_{t_n}))) \right) \quad (4.7.62)$$

$$\geq \sum_{n=0}^{N} \left(\tilde{F}(F^{-1}(x_n + F(D_{t_n}))) - \tilde{F}(a_n F^{-1}(x_n + F(D_{t_n}))) \right) \quad (4.7.63)$$

A lower bound for the last line of (4.7.60) is given by

$$\tilde{F}(x) - \tilde{F}(\bar{a}_x x) \geq \inf_{y \in [\bar{a}_x x, x]} f(y) \left| \int_{\bar{a}_x x}^{x} z \, dz \right| \quad (4.7.64)$$

$$= \frac{1}{2}\left(1 - \bar{a}_x^2\right) x^2 \inf_{y \in [\bar{a}_x x, x]} f(y) \geq 0. \quad (4.7.65)$$

Because of the assumptions (4.4.14) and (4.2.17), we know

$$H(x) := \frac{1}{2}\left(1 - \bar{a}_{F^{-1}(x)}^2\right)\left(F^{-1}(x)\right)^2 \inf_{y \in [\bar{a}_{F^{-1}(x)} F^{-1}(x), F^{-1}(x)]} f(y) \qquad (4.7.66)$$

tends to infinity for $|x| \to \infty$. Finally, we introduce the mapping

$$T(x) := (x_0, x_1 + F(D_{t_1}), \ldots, x_N + F(D_{t_N})), \qquad (4.7.67)$$

for which $C^{(2)}(x) \geq H(||T(x)||_\infty)$ holds. It remains to show that $||T(x)||_\infty \to \infty$ for $|x| \to \infty$. Let us assume there is a sequence x^k such that $||x^k||_\infty \to \infty$ but $||T(x^k)||_\infty$ remains bounded. This implies especially the boundedness of (x_0^k). But then again, $D_{t_1}^k = a_0^k F^{-1}(x_0^k)$ remains bounded. We can continue the argumentation for all coordinates of $T(x)$ and conclude that (x_n^k) is a bounded sequence for all $n \in \{0, \ldots, N\}$. This contradicts the assumption, thus, the lemma is proven. $\square$

Lemma 4.10. *Under the assumptions of Theorem 4.3, equation (4.4.16) has at most one solution, which is positive if existing. Furthermore, $g(x) := f(x) - \bar{a}_x f(\bar{a}_x x)(1 - \tau \bar{\rho}'(x)x)$ is positive.*

Proof. We show that both $h_2 \circ F^{-1}$ and

$$\hat{h}_2(x) := -F^{-1}\left(X_0 - N\left[x - F\left(\bar{a}_{F^{-1}(x)} F^{-1}(x)\right)\right]\right) \qquad (4.7.68)$$

are strictly increasing. In this case, at most one zero can exist, and its positivity is guaranteed by $h_2(F^{-1}(0)) = 0$ and $\hat{h}_2(0) = -F(X_0) < 0$. The function h_2 is strictly increasing because it is continuous, bijective, has a fixed point at zero and

$$\lim_{\epsilon \to 0} \frac{h_2(\epsilon) - h_2(0)}{\epsilon} = \lim_{\epsilon \to 0} \frac{f(\epsilon) - \bar{a}_\epsilon^2 f(\bar{a}_\epsilon \epsilon)(1 - \tau \bar{\rho}'(\epsilon)\epsilon)}{f(\epsilon) - \bar{a}_\epsilon f(\bar{a}_\epsilon \epsilon)(1 - \tau \bar{\rho}'(\epsilon)\epsilon)} \qquad (4.7.69)$$

$$= \frac{1 - \bar{a}_0^2}{1 - \bar{a}_0} > 0. \qquad (4.7.70)$$

Since F^{-1} is also strictly increasing, we have proven the same property for $h_2 \circ F^{-1}$.

4.7. PROOFS OF THE THEOREMS 4.3 AND 4.4

We differentiate $\hat{h}_2$:

$$\hat{h}_2'(x) = N\left[\frac{f(F^{-1}(x)) - \bar{a}_{F^{-1}(x)}f(\bar{a}_{F^{-1}(x)}F^{-1}(x))(1 - \tau\bar{\rho}'(F^{-1}(x))F^{-1}(x))}{f(F^{-1}(x))f\left(F^{-1}\left(X_0 - N\left[x - F(\bar{a}_{F^{-1}(x)}F^{-1}(x))\right]\right)\right)}\right]. \quad (4.7.71)$$

This expression is strictly positive, because the numerator is strictly positive as we show next. We define both

$$\begin{aligned} k(x) &:= f(x) - \bar{a}_x f(\bar{a}_x x)(1 - \tau\bar{\rho}'(x)x) \text{ and} \\ k_2(x) &:= f(x) - \bar{a}_x^2 f(\bar{a}_x x)(1 - \tau\bar{\rho}'(x)x). \end{aligned} \quad (4.7.72)$$

The numerator of (4.7.71) can be expressed by $k(F^{-1}(x))$, and furthermore, $h_2(x) = xk_2(x)/k(x)$. Both functions, k and k_2, are continuous, and due to the properties of h_2 explained in the beginning of the proof, the functions must have the same sign for all $x \in \mathbb{R}$. The function k_2 is strictly greater than k for all $x \in \mathbb{R}$; thus, there can be no change of signs and we have either $k(x) > 0$ and $k_2(x) > 0$ or $k(x) < 0$ and $k_2(x) < 0$ for all x. Because $k(0) = f(0)(1 - \bar{a}_0) > 0$, positivity is proven. $\square$

Lemma 4.11. *For all $n \in \{0, \ldots, N-1\}$, the partial derivatives of $C^{(1)}$ can be expressed by*

$$\frac{\partial}{\partial x_n}C^{(2)}(x) = D_{t_n+} + \frac{a_n f(D_{t_{n+1}})(1 - \tau\bar{\rho}'(D_{t_n+})D_{t_n+})}{f(D_{t_n+})}\left[\frac{\partial}{\partial x_{n+1}}C^{(2)}(x) - D_{t_{n+1}}\right]. \quad (4.7.73)$$

Proof. First, observe

$$\frac{\partial}{\partial x_n}D_{t_m} = \frac{a_n f(D_{t_{n+1}})}{f(D_{t_n+})}\left(1 - \tau\bar{\rho}'(D_{t_n+})D_{t_n+}\right)\frac{\partial}{\partial x_{n+1}}D_{t_m} \quad (4.7.74)$$

for $n \in \{0, \ldots, m-2\}$. This follows by expanding D_{t_m},

$$\frac{\partial}{\partial x_n}D_{t_m} = \frac{\partial}{\partial x_n}\left[a_{m-1}F^{-1}(x_{m-1} + F(\ldots(a_n F^{-1}(x_n + F(D_{t_n}))) \ldots))\right], \quad (4.7.75)$$

and applying the chain rule:

$$\frac{\partial}{\partial x_n} D_{t_m} = \left[\prod_{k=n+1}^{m-1} \left[\frac{d}{dx} \bar{a}_{F^{-1}(x_k+F(x))} F^{-1}(x_k+F(x)) \right]_{x=D_{t_k}} \right]$$
$$\cdot \left[\frac{\partial}{\partial x_n} \bar{a}_{F^{-1}(x_n+F(D_{t_n}))} F^{-1}(x_n+F(D_{t_n})) \right] \qquad (4.7.76)$$
$$= \left[f(D_{t_{n+1}}) \frac{\partial}{\partial x_{n+1}} D_{t_m} \right] \left[\frac{a_n (1 - \tau \bar{\rho}'(D_{t_n+}) D_{t_n+})}{f(D_{t_n+})} \right]. \qquad (4.7.77)$$

We use (4.7.74) and (4.7.58) for the transformation

$$\frac{\partial}{\partial x_n} C^{(2)}(x) \qquad (4.7.78)$$
$$= F^{-1}(x_n + F(D_{t_n})) + \sum_{m=n+1}^{N} \frac{\partial}{\partial x_n} \left[G(x_m + F(D_{t_m})) - \tilde{F}(D_{t_m}) \right] \qquad (4.7.79)$$
$$= D_{t_n+} + \sum_{m=n+1}^{N} f(D_{t_m}) \left[\frac{\partial}{\partial x_n} D_{t_m} \right] \left[F^{-1}(x_m + F(D_{t_m})) - D_{t_m} \right] \qquad (4.7.80)$$
$$= D_{t_n+} + \frac{a_n f(D_{t_{n+1}})}{f(D_{t_n+})} \left(1 - \tau \bar{\rho}'(D_{t_n+}) D_{t_n+}\right) \left(D_{t_{n+1}+} - D_{t_{n+1}}\right)$$
$$+ \sum_{m=n+2}^{N} f(D_{t_m}) \left[\frac{\partial}{\partial x_{n+1}} D_{t_m} \right] \left[F^{-1}(x_m + F(D_{t_m})) - D_{t_m} \right]. \qquad (4.7.81)$$

Now, the same calculation for $\partial C^{(2)}(x)/\partial x_{n+1}$ results in

$$\frac{\partial}{\partial x_n} D_{t_m} = D_{t_{n+1}+} + \sum_{m=n+2}^{N} f(D_{t_m}) \left[\frac{\partial}{\partial x_{n+1}} D_{t_m} \right] \left[F^{-1}(x_m + F(D_{t_m})) - D_{t_m} \right], \qquad (4.7.82)$$

and combining (4.7.81) and (4.7.82) yields the desired result. □

Finally, we are prepared to prove Theorem 4.3.

Lemma 4.12. *Strategy $\xi^{(2)}$ is the unique minimiser of function $C^{(2)}$, and all components of $\xi^{(2)}$ are positive.*

Proof. Lemma 4.9 guarantees the existence of at least one optimal strategy $\xi^* \in \Xi$. By standard arguments, there is a Lagrange multiplier $\nu \in \mathbb{R}$ such that

$$\frac{\partial}{\partial x_n^*} C^{(2)}(\xi^*) = \nu \quad \text{for } n \in \{0, \dots, N\}. \qquad (4.7.83)$$

4.7. PROOFS OF THE THEOREMS 4.3 AND 4.4

We use Lemma 4.11 to get

$$\nu = D_{t_n+} + \frac{a_n f(a_n D_{t_n+})}{f(D_{t_n+})}(1 - \tau \bar{\rho}'(D_{t_n+})D_{t_n+})\left[\nu - a_n D_{t_n+}\right] \quad (4.7.84)$$

$$\Leftrightarrow \nu = D_{t_n+} \frac{f(D_{t_n+}) - a_n^2 f(a_n D_{t_n+})(1 - \tau \bar{\rho}'(D_{t_n+})D_{t_n+})}{f(D_{t_n+}) - a_n f(a_n D_{t_n+})(1 - \tau \bar{\rho}'(D_{t_n+})D_{t_n+})} = h_2(D_{t_n+}) \quad (4.7.85)$$

for $n \in \{0, \ldots, N-1\}$. Function h_2 is one-to-one, and thus

$$\nu = h_2(F^{-1}(x_n^* + F(D_{t_n}))) \quad (4.7.86)$$

implies that $x_n^* + F(D_{t_n})$ does not depend on $n \in \{0, \ldots, N-1\}$. Consequently, $D_{t_n+} = F^{-1}(x_n^* + F(D_{t_n}))$ is constant in n such that we can conclude

$$\begin{aligned}
x_0^* &= F(h_2^{-1}(\nu)), & (4.7.87) \\
x_n^* &= x_0^* - F(D_{t_n}) \\
&= x_0^* - F(a_0 F^{-1}(x_0^*)) \text{ for } n \in \{1, \ldots, N-1\}, & (4.7.88) \\
x_N^* &= X_0 - x_0^* - (N-1)\left[x_0^* - F(a_0 F^{-1}(x_0^*))\right]. & (4.7.89)
\end{aligned}$$

The value x_0^* determines the optimal solution completely, and thus it must minimise

$$\begin{aligned}
& C_0^{(2)}(x_0) & (4.7.90) \\
:= & C^{(2)}\left(x_0, x_0 - F\left(a_0 F^{-1}(x_0)\right), \ldots, & (4.7.91) \\
& X_0 - x_0 - (N-1)\left[x_0 - F(a_0 F^{-1}(x_0))\right]\right) \\
\stackrel{(4.7.58)}{=} & G(x_0) + \sum_{n=1}^{N-1}\left[G(x_0) - \tilde{F}(a_0 F^{-1}(x_0))\right] & (4.7.92) \\
& + G(X_0 - N[x_0 - F(a_0 F^{-1}(x_0))]) - \tilde{F}(a_0 F^{-1}(x_0)) \\
= & N[G(x_0) - \tilde{F}(a_0 F^{-1}(x_0))] & (4.7.93) \\
& + G(X_0 - N[x_0 - F(a_0 F^{-1}(x_0))]).
\end{aligned}$$

Differentiation results in

$$\frac{dC_0^{(2)}(x_0)}{dx_0} = N\left[D_{0+} - a_0^2 D_{0+}\frac{f(D_{t_1})}{f(D_{0+})}\left(1 - \tau\bar{\rho}'(D_{0+})D_{0+}\right)\right.$$
$$\left. + D_{t_n+}\left(a_0\frac{f(D_{t_1})}{f(D_{0+})}\left(1 - \tau\bar{\rho}'(D_{0+})D_{0+}\right) - 1\right)\right] \quad (4.7.94)$$

such that we can calculate the minimiser by

$$\begin{aligned}\frac{d}{dx_0^*}C_0^{(2)}(x_0^*) &= 0 \\ \Leftrightarrow \quad D_{t_N+} &= D_{0+}\frac{f(D_{0+}) - a_0^2 f(D_{t_1})(1-\tau\bar{\rho}'(D_{0+})D_{0+})}{f(D_{0+}) - a_0 f(D_{t_1})(1-\tau\bar{\rho}'(D_{0+})D_{0+})}.\end{aligned} \quad (4.7.95)$$

The left-hand side of the last line can be rewritten as

$$D_{t_N+} = F^{-1}(F(D_{t_N}) + x_N^*) \quad (4.7.96)$$
$$= F^{-1}(F(D_{t_1}) + X_0 - x_0^* - (N-1)(x_0^* - F(D_{t_1}))) \quad (4.7.97)$$
$$= F^{-1}(X_0 - N(x_0^* - F(D_{t_1}))), \quad (4.7.98)$$

and the right-hand side is just $h_2(F^{-1}(x_0^*))$. We know by Lemma 4.10 that equation (4.7.95) has at most one zero such that we are finished with the existence, uniqueness and representation of the optimal strategy.

At last, we show that all components of this strategy are positive. We already know $x_0^* > 0$ by Lemma 4.10, and thus also $x_n^* > 0$ for all $n \in \{1, \ldots, N-1\}$ by (4.7.88). For the positivity of x_N^*, we transform (4.7.95) into

$$D_{t_N+} = D_{0+}\left[1 + \frac{a_0 f(a_0 D_{0+}) - a_0^2 f(a_0 D_{0+})}{f(D_{0+}) - a_0 f(a_0 D_{t_0+})(1 - \tau\bar{\rho}'(D_{0+})D_{0+})}\left(1 - \tau\bar{\rho}'(D_{0+})D_{0+}\right)\right]. \quad (4.7.99)$$

The fraction on the right-hand side is strictly positive by Lemma 4.10; positivity of x_N^* follows from

$$D_{t_N+} > D_{0+} = \frac{D_{t_N}}{a_0} > D_{t_N}. \quad (4.7.100)$$

$\square$

Appendix A

SimStocki - a simulation tool for the Opinion Game

The numerical results in our work have been produced with SIMSTOCKI[a], a *Java 1.5* based implementation of the Opinion Game algorithm. Not only provides SIMSTOCKI free portability to all common computer systems, it also has a graphical user interface (*GUI*) (see Figure A.1) to observe the time evolution of the system and change the parameter settings in realtime. In the stage of development, we incorporated several new parameters to see how they influence the dynamics.

We explain these new parameters next. They give an impression of the many ways in which the Opinion Game can be extended. Afterwards, we turn to the program itself. We give a summary of the classes[b], and present the most important ones in detail. We avoid auxiliary classes (writing and reading files, pure *data container*) and those parts of the program that are required for the graphical output. Although these parts would give an occasion to discuss some interesting issues concerning the graphics performance in Java, this topic goes beyond our scope here (and would add more than 50 pages of source code).

The implemented algorithm is part of Bovier et al. (2006). The source code is written by Alexander Weiß. All rights are reserved to the particular authors.

---

[a]We are grateful to Prof. Dr. Nils Berglund for the felicitous name.
[b]A comprehensive text book about Java and object oriented programming is Louis and Müller (2005).

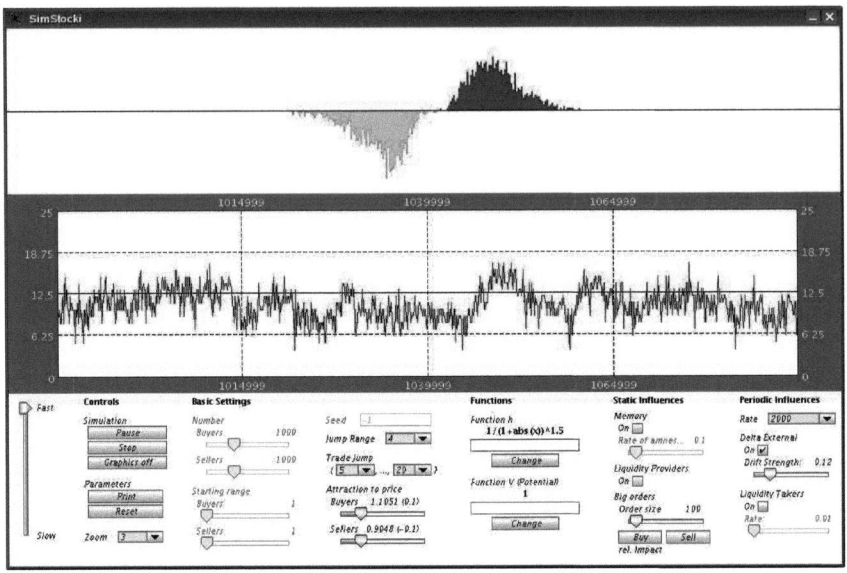

Figure A.1: Screenshot of SIMSTOCKI's GUI.

A.1 An exterior view

The GUI of SIMSTOCKI consists of three panels (see Figure A.1 again). The upper panel shows the dynamics of the system. The agents, respectively their opinions, are displayed as squares, the x-axis is the logarithmic price. Dark grey squares are sellers, light grey squares are buyers. The negative orientation of the buyers acts just as a second attribute to distinct buyers from sellers. The middle panel shows the generated price process. The axes of this graph are scaled dynamically. In the lower part of the GUI, buttons and sliders are provided to change the parameter settings.

In addition to the parameters of the original algorithm, explained in Section 2.2, and the extension for large orders from Section 4.3, there are some options that we have added later to the model:

- Memory

    This parameter is a first try to experiment with non-Markovian dynamics. In contrast to the original algorithm, each trader remembers the last price he or she has traded for, and is not willing to trade for a higher (in case of a buyer), respectively lower (seller), price. If the dynamics suggest a change of opinion to such a worse price, it is only accepted with probability p_{amn}, the *rate of amnesia*.

- Liquidity providers

  Bouchaud et al. (2004) distinguish two types of traders. Liquidity providers are traders whose

  > "[...] *profit comes from the bid-ask spread s: the sell price is always slightly larger than the buy price, so that each round turn operation leads to a profit equal to the spread s, at least if the midpoint has not changed in the mean time [...].*"

  A simple implementation of this idea in SIMSTOCKI forces the traders to jump to the best quotes after a trade.

- Liquidity takers

  Liquidity takers are the other considered group of traders. They

  > "[...] *trigger trades by putting in market orders. The motivation for this category of traders might be to take advantage of some 'information', and make a profit from correctly anticipating future price changes.*"

  Market orders are placed at the best quotes, and thus, they are executed immediately. As liquidity takers break their orders into smaller pieces to avoid negative price effects, they are closely connected to the discussion of large orders in Chapter 4. In SIMSTOCKI, they are modelled by a probability p_{liq}. With this probability, a chosen trader jumps directly to the best price of the other group, independent of his or her current position. In this way, a trade is forced.

The particular influences of these parameters on the evolution of the market are an interesting subject for future studies.

A.2 An interior view

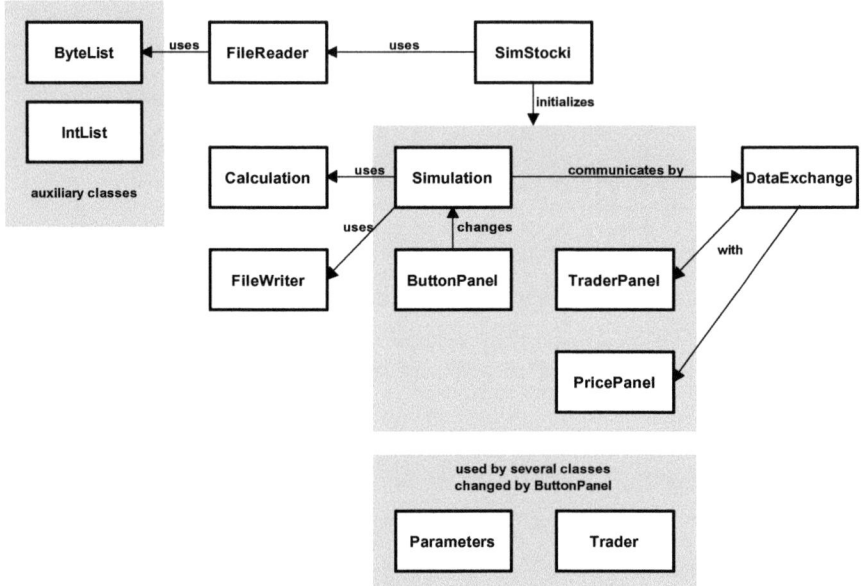

Figure A.2: Class dependencies in SIMSTOCKI.

The structure of SIMSTOCKI is illustrated in Figure A.2. The program is started by `SimStocki` that reads given parameters and initialises `Simulation`, `Parameters`, `Trader`, and the three parts of the graphical output `TraderPanel`, `PricePanel`, `ButtonPanel`. The class `Simulation` contains the main routine, which calls `Calculation` that does the particular updating procedure. The required parameter values are provided by `Parameters`, information about the current configuration are collected in `Trader`. The data that `Simulation` gets from `Calculation` are provided to the graphics classes. `TraderPanel` and `PricePanel` run as individual threads such that a slow graphics performance does not affect the speed of the main algorithm. The class `ButtonPanel` is able to change values in `Simulation`, `Parameters` and `Trader` to implement user input into the running simulation.

A.2. AN INTERIOR VIEW

In the following, we list the source code of **Simulation**, **Calculation** and **Trader**. We have removed the bodies of those methods that are trivial or not understandable without the missing classes. However, this reduction helps to focus the attention even more on the crucial parts of the programming code. Since, the code is well documented, we restrict ourselves to some short initial remarks about the single classes.

A.2.1 Simulation.java

The main method of this class is **run()** that starts in line 167. Basically, it consists of an infinite loop[c], which contains calls for the methods that communicate with the graphics classes (lines 203 and 204) and for the method **updateConfiguration()** (line 207).

The method **updateConfiguration()** begins in line 333. The first part (lines 336-352) consists of commands to write the generated data (prices, returns, gap size) into files. This part can easily be modified and extended. The rest of the method only calls **Calculation.calculateRound(...)**, but there is a case distinction to determine the parameter values that are given over. In the lines 353-372, we address the case that a large order is executed. Then, the fourth argument of **Calculation.calculateRound(...)** is **true**, and the fifth argument contains the information if the large order is a buy or a sell order. If there is a normal update, the lines 374-378 are applied.

```
1  package de.wiasberlin.simstocki;
2
3  import java.util.*;
4
5  /**
6   * This object is the nerve centre of SimStocki. While {@link Calculation}
7   * implements the algorithm itself, Simulation provides the interface for
8   * the other threads to communicate with Calculation. In particular,
9   * Simulation is a thread that calls Calculation in every round and
10  * communicates with the othere threads if necessary.
11  *
12  * @author Alexander Weiss
13  */
14
15 public class Simulation {
```

---

[c]If the user has chosen to let the simulation run a certain number of rounds only, the program cancels in the rough way (lines 255-264).

```
// ----------------------
// private class variables
// ----------------------

private static final int EDGELENGTH = 2;

// given classes

private Trader t;

private Parameters p;

private Random r;

private String dataFileName;

// recording data

private FileWriter fileReturn, filePrice, fileWindow;

private double oldPrice, newPrice;

private int currentPrice;

// counting ticks

private int ticks;

private int currTicks;

// measuring performance

private long startingTime;

private int percent;

// graphics

private boolean graphicsOn;

// data for Trader Panel

private int tpMinPrice, tpMaxPrice;

private int tpIntervalLength;

private int[] tpOrderedTraders;

private IntList tpFrom, tpTo, tpLabel;

private boolean tpRedrawAll;

// data for Price Panel

private int ppHistoryLength = 100000;

private int ppOverlap = ppHistoryLength / 10;

private int ppRounds;
```

```
 76
 77    private int ppArrayPos;
 78
 79    private long ppLastValue;
 80
 81    private int[] ppCurrentPriceHistory;
 82
 83    // influence by user;
 84    private int delayMilli = 0;
 85
 86    private int delayNano = 0;
 87
 88    private boolean isPaused = false;
 89
 90    private boolean isStopped;
 91
 92    private boolean isStoppedAtFirst;
 93
 94    private boolean priceHistoryLengthChanged;
 95
 96    private int tempIndex;
 97
 98    // large orders
 99
100    // this variable gets only attention if a large order is being executed
101    private boolean largeOrderJustStarted;
102
103    private int largeOrderVolume;
104
105    // ---------------------
106    // public class variables
107    // ---------------------
108
109    public DataExchange dataForTP = new DataExchange();
110
111    public DataExchange dataForPP = new DataExchange();
112
113    public DataExchange dataForBP = new DataExchange();
114
115    // -----------
116    // public part
117    // -----------
118
119    /**
120     * Initialises itself, but also TraderPanel and PricePanel.
121     *
122     * @param dataFileName
123     *            name for files containing recorded data
124     * @param startOn
125     *            if true, simulation starts immediately
126     * @param graphicsOn
127     *            if true, evolution is visible on screen
128     * @param ticks
129     *            running time
130     */
131    public Simulation(Trader t, Parameters p, String dataFileName,
132        int panelWidth, boolean startOn, boolean graphicsOn, int ticks) {...}
133
134    /**
135     * Closes the data files when the program is finished or stopped.
```

```java
     *
     */
    public void closeFiles() {...}

    /**
     * Checks if graphical output is switched on.
     *
     * @return true if graphical output is switched on
     */
    public boolean graphicsOn() {...}

    /**
     * Triggers if graphical output is switched on.
     *
     * @param value
     *              if true, graphics on; else off
     */
    public void graphicsOn(boolean value) {...}

    /**
     * Checks if simulation is running.
     *
     * @return true if it is stopped, false if it is running
     */
    public boolean isStopped() {...}

    /**
     * The main method of this class, controlling the algorithm and updating
     * the panels.
     *
     */
    public void run() {

      // it runs forever (or is exited in a controlled way)
      while (true) {

        // if the program is stopped, we do nothing...
        if (isStopped) {

          // ... only if it has been stopped only recently, we have to set
          // back the settings
          if (isStoppedAtFirst) {
            closeFiles();
            p.reset();
            t.reset(p);
            initialize();
            updateTraderPanel(true);
            synchronized (dataForPP) {
              dataForPP.reset = true;
              dataForPP.notifyAll();
            }
            r = p.r();
            isStoppedAtFirst = false;
          }
        }

        // ...else the simulation is being executed
        else {

          // record starting time to evaluate performance
```

A.2. AN INTERIOR VIEW

```
196          if (currTicks == 0)
197            startingTime = System.currentTimeMillis();
198
199          // if final number of ticks has not been reached yet
200          if (ticks == 0 || currTicks < ticks) {
201
202            // update the panels
203            updateTraderPanel(false);
204            updatePricePanel();
205
206            // update the configuration
207            updateConfiguration();
208
209            // count the number of simulation steps
210            currTicks++;
211
212            // change length of price history
213            if (priceHistoryLengthChanged)
214              setPriceHistoryLength();
215
216            try {
217
218              // if simulation is paused ....
219              if (isPaused) {
220                synchronized (dataForBP) {
221
222                  // it sleeps until it is woken up by ButtonPanel
223                  dataForBP.wait();
224                  isPaused = false;
225                }
226              }
227
228              // the delay in the simulation
229              Thread.sleep(delayMilli, delayNano);
230
231            } catch (InterruptedException e) {
232            }
233
234            // calculating progress and estimating remaining time (if
235            // tick number is finite)
236            if (ticks != 0) {
237              if (currTicks * 100. / ticks >= percent) {
238                System.out.println();
239                System.out.println(percent + "% done");
240                System.out
241                  .println("Remaining time: "
242                      + (int) (((System
243                          .currentTimeMillis() - startingTime)
244                          * (100. - percent) / percent) / 60000)
245                      + " minutes");
246                percent++;
247              }
248            }
249          }
250
251          // final number of ticks achieved
252          else {
253
254            // calculating overall performance
255            System.out.println();
```

```java
            System.out
                .println("Simulation done in "
                    + (int) ((System.currentTimeMillis() - startingTime) /
                        60000.)
                    + " minutes with " + currTicks + " steps.");

            // tidying up and exiting
            closeFiles();
            System.exit(0);
        }
      }
    }
}

/**
 * Let the simulation (re)run if it is stopped.
 *
 */
public void runSimulation() {...}

/**
 * Governs the speed of the simulation. The larger the delay, the slower
 * the simulation.
 *
 * @param value
 *            delay in nanoseconds
 */
public void setDelay(int value) {...}

/**
 * Let the simulation pause.
 *
 */
public void setPaused() {...}

/**
 * Sets a particular boolean to true such that {@link run} calls
 * {@link setPriceHistoryLength} to update the length of the displayed
 * price history.
 *
 * @param index
 *            length of history: 0 = 10000; 1 = 50000; 2 = 100000; 3 =
 *            500000
 */
public void changePriceHistoryLength(int index) {...}

/**
 * Stops the simulation and resets the parameters.
 *
 * @param value
 *            if true a text message about the reset is written to the
 *            terminal
 */
public void setStopped(boolean value) {...}


// ------------
// private part
// ------------

```

A.2. AN INTERIOR VIEW

```
/**
 * Part of the constructor, but also used when the system is reset by user
 *
 */
private void initialize() {...}

/**
 * Is called when a new interval length of the price chart is chosen by
 * the user.
 *
 */
private void setPriceHistoryLength() {...}

/**
 * Calls {@link Calculation} and writes data to the files. Best place to
 * introduce new commands for collecting data.
 *
 */
private void updateConfiguration() {

  // we record the data all 100 ticks
  if (currTicks % 100 == 0) {

    newPrice = t.getPrice();

    // the returns
    if (currTicks != 0)
      fileReturn.write(newPrice - oldPrice);

    // the price
    filePrice.write(newPrice);

    // the gap size
    fileWindow.write(t.getAskPrice() - t.getBidPrice());

    oldPrice = newPrice;
  }

  // if there is a large order, ...
  if (p.isLargeOrderExecuting()) {
    if (largeOrderJustStarted) {
      largeOrderJustStarted = false;
      largeOrderVolume = p.getLargeOrderSize();
    }

    // ... we force the algorithm to trade a share in this round, ...
    Calculation.calculateRound(t, p, r, true, p.isLargeOrderBought(),
        tpFrom, tpTo, tpLabel);

    // ... and decrease the volume of the large order by 1, ...
    largeOrderVolume--;

    // ... until the whole volume is traded
    if (largeOrderVolume == 0) {
      p.stopLargeOrder();
      largeOrderJustStarted = true;
    }
  }
```

```
374      // else, we just apply the standard algorithm
375      else {
376        Calculation.calculateRound(t, p, r, false, false, tpFrom, tpTo,
377             tpLabel);
378      }
379    }
380
381    /**
382     * Synchronises data and notifies PricePanel.
383     *
384     */
385    private void updatePricePanel() {...}
386
387    /**
388     * Synchronises data and notifies TraderPanel.
389     *
390     * @param firstRound
391     *              if true, a graphical output is forced (after the start)
392     */
393    private void updateTraderPanel(boolean firstRound) {...}
394  }
```

A.2.2 Calculation.java

The main method of Calculation.java is calculateRound(...). It contains the basic algorithm and begins in line 37. Until line 118, the given parameter values are considered to determine which trader is updated and to what value his or her opinion is updated to. If there is no large order executed, the trader is chosen in line 75 by calling the method chooseTrader(...), and if a special way of updating (large order, liquidity takers, ...) is not chosen, the difference between old and new opinion is determined in line 104 by calling the method chooseStep(...). In the lines 134 to 210, we consider the case that the chosen trader buys a share: A trading partner must be found (137-139), the share must be exchanged and the new opinions must be determined (142-200). The exchange itself is done by Trader.setOpinions(...) in the lines 202-205. We see in the next section that this method changes the configuration and refreshes whatever needs to be refreshed (list of traders at the best prices, the price itself, ...). We treat the *symmetric* case that the chosen trader sells a share in the lines 214-254, and the case that no trade happens in the lines 257-271.

The method chooseTrader(...) (368-410) implements the function g, defined in line (2.2.1) on page 16. The method chooseStep(...) (286-355) implements function f as given in line (2.2.4) of the Opinion Game algorithm on page 16.

A.2. AN INTERIOR VIEW

```java
package de.wiasberlin.simstocki;

import java.util.*;

/**
 * Determines the next change of the configuration. The core class of the
 * whole program.
 *
 * @author Alexander Weiss
 *
 */
public class Calculation {

  // -----------------------
  // private class variables
  // -----------------------

  static int currentTrader;

  static int step;

  // number of traders sitting at the best prices
  static int numBestSellers, numBestBuyers;

  // -----------
  // public part
  // -----------

  /**
   * Performs a change in the opinion of one trader. The trader is chosen
   * and the size of his/her opinion change is determined. A little bit
   * complicated, since all kinds of special trading behaviour (large
   * orders, liquidity takers, liquidity providers, ...) must be checked.
   *
   * @return true if a trade event happened, false else.
   */
  public static boolean calculateRound(Trader t, Parameters p, Random r,
      boolean largeOrder, boolean largeOrderBought, IntList from,
      IntList to, IntList label) {

    boolean hasTradeHappened = false;

    boolean isLiqTaker;

    boolean isBuyer;

    p.updateExternalInfluence();

    /*
     * first, we determine the index of the trader whose opinion will be
     * changed
     */

    // if a large order is executed, we choose a trader from the end of the
    // tail
    if (largeOrder) {
      currentTrader = 0;

      // if large order is buy order, look for lowest opinion
      if (largeOrderBought) {
        for (int i = 0; i < p.getNumTraders(); i++)
```

```
            if (t.getOpinion(currentTrader) > t.getOpinion(i))
              currentTrader = i;
        }

        // else look for highest opinion
        else
          for (int i = 0; i < p.getNumTraders(); i++)
            if (t.getOpinion(currentTrader) < t.getOpinion(i))
              currentTrader = i;
      }

      // if no large order is executed, use standard algorithm
      else
        currentTrader = chooseTrader(t, p, r);

      // determine characteristics of chosen trader
      isBuyer = t.isBuyer(currentTrader);
      isLiqTaker = p.isLiqTaker(isBuyer);

      /*
       * next, we determine the step size of his/her opinion change
       */

      // if a large order is executed, the trader jumps directly to the best
      // price of the other trader group
      if (largeOrder)
        if (largeOrderBought)
          step = t.getAskPrice() - t.getOpinion(currentTrader);
        else
          step = t.getBidPrice() - t.getOpinion(currentTrader);

      // similarly, the trader can be a liquidity taker
      else if (isLiqTaker) {
        if (isBuyer)
          step = t.getAskPrice() - t.getOpinion(currentTrader);
        else
          step = t.getBidPrice() - t.getOpinion(currentTrader);

      }

      // else determine the step by the standard algorithm
      else
        step = chooseStep(t, p, r, t.getOpinion(currentTrader));

      // check if trader is not willing to perform the change of opinion,
      // since he/she would lose money when trading for this price (switched
      // off while large order is performed)
      if (!largeOrder && p.hasMemory())
        if (t.remembersLastTradePrice(currentTrader, step))

          // finally, he/she could forget about the old price
          if (p.hasAmnesia())
            t.forgetLastTradePrice(currentTrader);

          // if not, there is no change in opinion
          else
            step = 0;

      /*
       * we perform the change of opinion
```

```
        */


        // nothing to do if there's no change
        if (step != 0)

            // is the new configuration unstable such that a trade happens?
            if (t.isTradeEvent(currentTrader, t.getOpinion(currentTrader)
                + step)) {

                hasTradeHappened = true;

                // again a distinction in buyers ...
                if (isBuyer) {

                    // trading partner randomly chosen from best price
                    numBestSellers = t.getBestSellers().size();
                    int otherTrader = t.getBestSellers().get(
                        r.nextInt(numBestSellers));

                    // step done by the trading partner
                    int otherStep;

                    // if there is a trade, the traders jump a away from trading
                    // price. already contained in otherStep for trading partner,
                    // but extra step must be determined for original trader.
                    int tradeStep = 0;

                    // if trading partner is a liquidity provider, he/she just
                    // jumps to the best price of the other group
                    if (p.containsLiqProvs())
                        otherStep = t.getBidPrice() - t.getAskPrice();

                    // else step size is determined by original method
                    else
                        otherStep = -p.getTradeJump();

                    // information for the price panel. the price panel only
                    // redraws parts of the configuration that have changed. To
                    // identify these parts, every move of a trader is recorded
                    // in three lists. "from" keeps the former opinion, "to"
                    // keeps the new opinion, and "label" keeps the (possible)
                    // change of a trader:
                    // 0: buyer -> buyer
                    // 1: buyer -> seller
                    // 2: seller -> buyer
                    // 3: seller -> seller
                    from.add(t.getOpinion(currentTrader));
                    from.add(t.getOpinion(otherTrader));
                    label.add(1);
                    label.add(2);

                    /*
                     * determine value of tradeStep
                     */

                    // if a large order is executed
                    if (largeOrder) {

                        // sample traders until you have found a seller
                        boolean found = false;
```

```
            while (!found) {
              int ranTrader = r.nextInt(p.getNumTraders());
              if (!t.isBuyer(ranTrader)) {
                found = true;

                // let current trader jump to the same opinion
                // as the sampled trader has => the distribution
                // of the final new opinion is proportional to
                // the shape of the group of sellers
                tradeStep = t.getOpinion(ranTrader)
                    - t.getAskPrice();
              }
            }
          }

          // else extra jump after trade is determined by original
          // method
          else
            tradeStep = p.getTradeJump();

          t.setOpinions(currentTrader, t.getOpinion(currentTrader)
              + step + tradeStep, otherTrader, t
              .getOpinion(otherTrader)
              + otherStep);

          // again data for the price panel (see line 158 etc.)
          to.add(t.getOpinion(currentTrader));
          to.add(t.getOpinion(otherTrader));
        }

        // ... and sellers. works equivalently to case of buyers. look
        // above for comments.
        else {
          numBestBuyers = t.getBestBuyers().size();
          int otherTrader = t.getBestBuyers().get(
              r.nextInt(numBestBuyers));

          int otherStep;

          int tradeStep = 0;

          if (p.containsLiqProvs())
            otherStep = t.getAskPrice() - t.getBidPrice();
          else
            otherStep = p.getTradeJump();

          from.add(t.getOpinion(currentTrader));
          from.add(t.getOpinion(otherTrader));
          label.add(2);
          label.add(1);

          if (largeOrder) {
            boolean found = false;
            while (!found) {
              int ranTrader = r.nextInt(p.getNumTraders());
              if (t.isBuyer(ranTrader)) {
                found = true;
                tradeStep = t.getOpinion(ranTrader)
                    - t.getBidPrice();
              }
```

A.2. AN INTERIOR VIEW

```
          }
        } else
          tradeStep = -p.getTradeJump();

        t.setOpinions(currentTrader, t.getOpinion(currentTrader)
            + step + tradeStep, otherTrader, t
            .getOpinion(otherTrader)
            + otherStep);

        to.add(t.getOpinion(currentTrader));
        to.add(t.getOpinion(otherTrader));
      }
    }

    // without a trade, we only update one opinion
    else {

      // information for the price panel (see line 158 etc.)
      from.add(t.getOpinion(currentTrader));
      if (t.isBuyer(currentTrader))
        label.add(0);
      else
        label.add(3);

      // update opinion
      t.setOpinion(currentTrader, t.getOpinion(currentTrader) + step);

      // information for the price panel
      to.add(t.getOpinion(currentTrader));
    }

    return hasTradeHappened;
  }

  // ------------
  // private part
  // ------------

  /**
   * Chooses the step (= change of opinion) the trader performs in this
   * round.
   *
   * @return size of the step.
   */
  private static int chooseStep(Trader t, Parameters p, Random r,
      int currentOpinion) {

    // we need this value several times in this method
    final double EXP = Math.exp(1);

    int maxStepSize = p.getJumpRange();

    // the probability of every step size to be chosen
    double[] probability = new double[maxStepSize * 2 + 1];

    double deltaBuyer = p.getDeltaBuyers();

    double deltaSeller = p.getDeltaSellers();

    double deltaExternal = p.getDeltaExternal();
```

```java
    double randomNum = r.nextDouble();

    // cumulated probability
    double cumProb = 0;

    // captures part of probability that is given by the potential variable
    double potPart;

    double potDiff; // potential difference, auxiliary variable

    for (int stepSize = -maxStepSize; stepSize <= maxStepSize; stepSize++) {

      // first we consider the probabilities that the step size is unequal
      // to 0
      if (stepSize != 0) {

        // calculate exp(V(p) - V(p+d)), first: V(p) - V(p+d)
        potDiff = p.V(currentOpinion) - p.V(currentOpinion + stepSize);

        // if extDiff is an integer, we trust our own method
        if ((int) potDiff - potDiff == 0)
          potPart = power(EXP, (int) potDiff);

        // else we use the standard java method
        else
          potPart = Math.exp(potDiff);

        // for buyers ...
        if (t.isBuyer(currentTrader))

          // ... the probability for the value stepSize is given by
          // (normalising term) * ((delta_b * deltaExt)^d * potPart)
          // "min" 1)
          probability[stepSize + maxStepSize] = (1. / (maxStepSize * 2 + 1))
              * Math.min(power(deltaBuyer * deltaExternal,
                  stepSize)
                  * potPart, 1);

        // for sellers ...
        else

          // ... it's the same with delta_s instead of delta_b
          probability[stepSize + maxStepSize] = (1. / (maxStepSize * 2 + 1))
              * Math.min(power(deltaSeller * deltaExternal,
                  stepSize)
                  * potPart, 1);

        cumProb += probability[stepSize + maxStepSize];

        if (randomNum < cumProb)
          return stepSize;
      }
    }

    // if step size is not unequal 0, it must be 0
    return 0;
  }

  /**
```

A.2. AN INTERIOR VIEW

143

```java
     * Chooses the trader whose opinion is changed in the current step. The
     * probability to be chosen depends on the trader's distance to the best
     * price.
     *
     * @return index of chosen trader
     */
    private static int chooseTrader(Trader t, Parameters p, Random r) {

      int numTraders = p.getNumTraders();

      // random number uniformly chosen in [0,1)
      double randomNum = r.nextDouble();

      double[] probability = new double[numTraders];

      double partFunc = 0; // partition function to normalise probabilities

      double cumProb = 0; // cumulated probability

      // calculate h values for the distance of each trader to the best price
      for (int i = 0; i < numTraders; i++) {

        // for buyers consider the distance to the bid price
        if (t.isBuyer(i))
          probability[i] = p.h(t.getBidPrice() - t.getOpinion(i));

        // for sellers consider the distance to the ask price
        else
          probability[i] = p.h(t.getOpinion(i) - t.getAskPrice());

        // by now, probabilities are no probabilities, because they are not
        // normalised. changes when divided by partition function.
        partFunc += probability[i];
      }

      // the probabilities to be chosen divide the interval [0,1) into
      // disjoint subintervals. look in which subinterval the random number
      // lies and return the index of the corresponding trader.
      for (int i = 0; i < numTraders; i++) {
        cumProb += probability[i];
        if (randomNum < cumProb / partFunc)
          return i;
      }

      // just to be sure that the program does not crash, return 0 if the
      // former algorithm did not succeed.
      System.out.println("Rounding error!");
      return 0;
    }

    /**
     * Calculates "base to the exponent" for integer values of the exponent.
     * Used, since we do not know if the particular standard method is
     * optimised.
     *
     * @return base^exponent
     */
    private static double power(double base, int exponent) {

      double retValue = 1;
```

```
    if (exponent == 0)
      return retValue;

    for (int i = 0; i < Math.abs(exponent); i++) {
      retValue *= base;
    }

    if (exponent < 0)
      retValue = 1 / retValue;

    return retValue;
  }
}
```

A.2.3 Trader.java

The class **Trader** provides many methods that just return certain variable values, for example the price, the status of a trader (buyer/seller) or a list of all traders with opinions at the best prices. The more sophisticated methods of **Trader** are setOpinion(...) (142-227) and setOpinions(...) (245-350). The former method is called by **Calculation** if the opinion of a single trader is updated, the latter one is called if a trade happens such that two opinions must be changed. The challenge in both methods is the performance. Since the program needs to choose a trader at the best quotes regularly, it is reasonable to register these traders in a list. Yet, the change of opinions can influence the list entries. A naive approach would be to rebuild the list every time an opinion is updated. However, this procedure is time consuming, as one has to check the position of every single trader. A more efficient implementation conserves the existing list as long as possible. The price for this increase in performance is a lengthy code with several case distinctions.

```
package de.wiasberlin.simstocki;

import java.util.*;

/**
 * Administrates state of traders, i.e. the configuration, and changes in
 * the configuration. There are methods to determine the (best) price(s),
 * to determine the traders sitting at these prices and to calculate
 * the changes if a trader changes his/her opinion.
 *
 * @author Alexander Weiss
 *
 */
```

A.2. AN INTERIOR VIEW

```
14  public class Trader {
15
16      // ------------------------
17      // private class variables
18      // ------------------------
19
20      // number of evaluated function values in buffer. could also be given by
21      // user when starting program but it's not so important. The standard
22      // window shows the area from {-250, ..., 250} in the TraderPanel. we have
23      // a buffer of 750 in each direction when bufferSize is 2001. bufferSize
24      // must be odd and and consistent with bufferSize value in class
25      // Parameters!
26      private static final int BUFFERSIZE = 2001;
27
28      // in the beginning, buyers and sellers are distributed around this price.
29      private static final int STARTPRICE = 0;
30
31      // in the beginning, sellers are uniformly distributed in [STARTPRICE + 1,
32      // STARTPRICE + rangeSellers + 1]. the buyers are distributed equally.
33      private int rangeBuyers, rangeSellers;
34
35      // as in class Parameters, there are standard values for certain
36      // variables. the same applies to the start ranges of the traders.
37      private int stdRangeBuyers, stdRangeSellers;
38
39      // the particular state of every trader. opinion contains the current
40      // opinion about the price, stocks is the number of possesed stocks (in
41      // this implementation given by 0 or 1), and lastTradePrice contains the
42      // price the particular trader has traded for last.
43      private int[] opinion, stocks, lastTradePrice;
44
45      // the number of traders and stocks given by class Parameters
46      private int numTraders, numStocks;
47
48      private int bidPrice, askPrice;
49
50      private double price;
51
52      // contains traders sitting on the best bid price
53      private IntList bestBuyer;
54
55      // contains traders sitting on the best ask price
56      private IntList bestSeller;
57
58      // -----------
59      // public part
60      // -----------
61
62      /**
63       * Initialises the class. Checks for validity of trader ranges and
64       * initialises the particular variables.
65       *
66       */
67      public Trader(Parameters p, int rangeBuyers, int rangeSellers) {...}
68
69      /**
70       * Deletes memory of the particular trader about the last price he/she
71       * traded for
72       *
73       * @param index
```

```java
 74     *              index of particular trader
 75     */
 76    public void forgetLastTradePrice(int index) {...}
 77
 78    public int getAskPrice() {...}
 79    public int getBidPrice() {...}
 80    public IntList getBestBuyers() {...}
 81    public IntList getBestSellers() {...}
 82    public double getPrice() {...}
 83    public int getRangeBuyers() {...}
 84    public int getRangeSellers() {...}
 85    public int getStdRangeBuyers() {...}
 86    public int getStdRangeSellers() {...}
 87    public int getOpinion(int index) {...}
 88    public int[] getOpinions() {...}
 89
 90    /**
 91     * Checks if particular trader is a buyer.
 92     *
 93     * @param index
 94     *              index of particular trader.
 95     * @return true if trader is buyer, false else.
 96     */
 97    public boolean isBuyer(int index) {...}
 98
 99    /**
100     * Checks if a trader causes a trade if he/she changes his/her opinon to a
101     * new value.
102     *
103     * @param index
104     *              index of the trader.
105     * @param opinion
106     *              Value of the new opinion
107     * @return true if a trade caused, false else.
108     */
109    public boolean isTradeEvent(int index, int value) {...}
110
111    /**
112     * Checks if a trader gets in conflict with his/her memory about the last
113     * price he/she traded for when he/she changes his/her opinion to the
114     * current value plus step.
115     *
116     * @param index
117     *              index of particular trader.
118     * @return true if trader gets in conflict, false else.
119     */
120    public boolean remembersLastTradePrice(int index, int step) {...}
121
122    /**
123     * Resets the variables of the class. Is called by the constructor and if
124     * system is reseted by user. Should be called after reset of class
125     * Parameters, as some values are taken from this class.
126     *
127     * @param p
128     *              instance of class parameter.
129     */
130    public void reset(Parameters p) {...}
131
132    /**
133     * Sets the opinion of a particular trader to a new value. Assumes that
```

```
     * the new configuration is stable, i.e. there is no trade necessary. This
     * assumption can be checked before with method isTradeEvent().
     *
     * @param index
     *              index of the particular trader.
     * @param value
     *              value of the new opinion.
     */
    public void setOpinion(int index, int value) {
      boolean wasBestTrader = false;

      // you cannot escape the range of the buffer
      if (value < -(BUFFERSIZE - 1) / 2)
        value = -(BUFFERSIZE - 1) / 2;
      if (value > (BUFFERSIZE - 1) / 2)
        value = (BUFFERSIZE - 1) / 2;

      // if the new opinion is not equal to the old one?
      if (opinion[index] != value) {

        // determine if trader is sitting on best price at the moment
        if (stocks[index] == 0 && opinion[index] == bidPrice)
          wasBestTrader = true;
        else if (stocks[index] == 1 && opinion[index] == askPrice)
          wasBestTrader = true;

        // change value
        opinion[index] = value;

        // if trader is buyer
        if (stocks[index] == 0) {

          // 1.) does new opinion cause a new best price
          if (value > bidPrice) {

            // there is a new best bid price and the particular trader
            // is the only trader sitting on it
            bestBuyer.clear();
            bestBuyer.add(index);
            bidPrice = value;
          }

          // 2.) does trader's new opinion match the bid price
          else if (value == bidPrice)

            // trader is added to the list of best buyers
            bestBuyer.add(index);

          // 3.) did trader belong to traders sitting on the best bid
          // price and does not anymore
          else if (wasBestTrader) {

            // remove trader from list
            bestBuyer.remove(index);

            // if list is empty now, rebuild it
            if (bestBuyer.isEmpty())
              reconstructBestBuyers();
          }
        }
```

```
      // if trader is seller
      else {

        // 1.) does new opinion cause new best price
        if (value < askPrice) {

          // same procedure as above
          bestSeller.clear();
          bestSeller.add(index);
          askPrice = value;
        }

        // 2.) does trader's new opinion match the bid price
        else if (value == askPrice)

          // same procedure as above
          bestSeller.add(index);

        // 3.) did trader belong to traders sitting on the best ask
        // price and does not anymore
        else if (wasBestTrader) {

          // same procedure as above
          bestSeller.remove(index);
          if (bestSeller.isEmpty())
            reconstructBestSellers();
        }
      }

      // update the price
      price = (bidPrice + askPrice) / 2.;
    }
  }

  /**
   * Sets the opinion of two particular traders to new values. Assumes that
   * the both traders are performing a trade with each other to get back
   * into a stable configuration. This assumption can be checked before with
   * method isTradeEvent(). Assumes furthermore that the trading price is
   * the price the second trader is sitting at (important for the memory).
   *
   * @param index1
   *          index of first particular trader.
   * @param value1
   *          value of new opinion of first particular trader.
   * @param index2
   *          index of second particular trader.
   * @param value2
   *          value of new opinion of second particular trader.
   */
  public void setOpinions(int index1, int value1, int index2, int value2) {
    int dummy;
    int oldValue1, oldValue2;

    lastTradePrice[index1] = opinion[index2];
    lastTradePrice[index2] = opinion[index2];

    // you cannot escape the range of the buffer
    if (value1 < -(BUFFERSIZE - 1) / 2)
```

A.2. AN INTERIOR VIEW

```
254       value1 = -(BUFFERSIZE - 1) / 2;
255     if (value2 < -(BUFFERSIZE - 1) / 2)
256       value2 = -(BUFFERSIZE - 1) / 2;
257     if (value1 > (BUFFERSIZE - 1) / 2)
258       value1 = (BUFFERSIZE - 1) / 2;
259     if (value2 > (BUFFERSIZE - 1) / 2)
260       value2 = (BUFFERSIZE - 1) / 2;
261
262     // update memory
263     lastTradePrice[index1] = opinion[index2];
264     lastTradePrice[index2] = opinion[index2];
265
266     // to simplify things, index1 is the buyer, and index2 is the trader
267     if (stocks[index2] == 0) {
268       dummy = index1;
269       index1 = index2;
270       index2 = dummy;
271       dummy = value1;
272       value1 = value2;
273       value2 = dummy;
274     }
275
276     // keep old opinions in mind before you update
277     oldValue1 = opinion[index1];
278     oldValue2 = opinion[index2];
279
280     // buyer becomes seller
281     stocks[index1] = 1;
282     opinion[index1] = value1;
283
284     // seller becomes buyer
285     stocks[index2] = 0;
286     opinion[index2] = value2;
287
288     // concerning the new best bid price
289
290     // 1.) new buyer's opinion is higher than old bid price
291     if (opinion[index2] > bidPrice) {
292
293       // new list with new buyer as only entry
294       bestBuyer.clear();
295       bestBuyer.add(index2);
296       bidPrice = opinion[index2];
297     }
298
299     // 2.) new buyer's opinion equals bid price
300     else if (opinion[index2] == bidPrice) {
301
302       // add new buyer to list
303       bestBuyer.add(index2);
304
305       // remove old buyer from list if necessary
306       if (oldValue1 == bidPrice)
307         bestBuyer.remove(index1);
308     }
309
310     // 3.) new buyer's opinion is lower than bid price
311     // remove old buyer from list and reconstruct list (if necessary)
312     else if (oldValue1 == bidPrice) {
313       bestBuyer.remove(index1);
```

```java
      if (bestBuyer.isEmpty())
        reconstructBestBuyers();
    }

    // concerning the new best ask price

    // 1.) new seller's opinion is lower than old ask price
    if (opinion[index1] < askPrice) {

      // new list with new seller as only entry
      bestSeller.clear();
      bestSeller.add(index1);
      askPrice = opinion[index1];
    }

    // 2.) new seller's opinion equals ask price
    else if (opinion[index1] == askPrice) {

      // add new buyer to list
      bestSeller.add(index1);

      // remove old seller from list if necessary
      if (oldValue2 == askPrice)
        bestSeller.remove(index2);
    }

    // 3.) new seller's opinion is higher than ask price
    // remove old seller from list and reconstruct list (if necessary)
    else if (oldValue2 == askPrice) {
      bestSeller.remove(index2);
      if (bestSeller.isEmpty())
        reconstructBestSellers();
    }

    // finally update the price
    price = (bidPrice + askPrice) / 2.;
  }

  public void setRangeBuyers(int value) {...}
  public void setRangeSellers(int value) {...}

  // ------------
  // private part
  // ------------

  /**
   * Determines the best bid price and constructs a list with all buyers
   * sitting on this price. Necessary if the best bid price has changed.
   */
  private void reconstructBestBuyers() {...}

  /**
   * Determines the best ask price and constructs a list with all sellers
   * sitting on this price. Necessary if the best ask price has changed.
   */
  private void reconstructBestSellers() {...}

}
```

Appendix B

Index of symbols

$\succ$		absolute continuity of measures: $\Psi \succ \varphi \;:\Leftrightarrow\; (\Psi(A)=0 \;\Rightarrow\; \varphi'(A)=0)$
$\cap$		intersection of sets
$\wedge$		logical operator "*and*"
$\cup$		union of sets
$\vee$		logical operator "*or*", also max-operator
$\|(\cdot,\cdot)\|_1$		1-norm: sum of the two coordinates' absolute values (in $\mathbb{R} \times \mathbb{R}$)
$\|\cdot\|_\infty$		sup-norm: maximum of absolute values of a vector's coordinates
a.s.		almost surely, with probability 1
$\mathfrak{B}(\cdot)$		Borel-σ-algebra
$\stackrel{d}{=}$		equality in distribution of two random variables
$\delta_\mathbf{x}$		Dirac measure for $x \in \mathbb{R}$
$\mathbb{E}(\cdot)$		mean of a random variable
$\mathbb{E}_{(x,y)}(\cdot)$		mean of a random variable induced by a process that is started in (x,y)
$\mathbb{N}$		the set of natural numbers
$\mathbb{N}_0$		$\mathbb{N} \cup \{0\}$
$\mathcal{O}$		Landau symbol: $f \in \mathcal{O}(g) \;:\Leftrightarrow\; (\exists c > 0)(\exists x_0)(\forall x > x_0)\; |f(x)| \leq c|g(x)|$
$\mathbb{R}$		the set of real numbers
sgn$(\cdot)$		sign function: -1 for negative arguments, 1 for positive ones, 0 for the argument 0
Θ		Landau symbol: $f \in \Theta(g) \;:\Leftrightarrow\; (\exists c^-, c^+ > 0)(\exists x_0)(\forall x > x_0)\; c^-|g(x)| \leq |f(x)| \leq c^+|g(x)|$
$t-$		as index of a function: left hand limit of the function at time t
$t+$		as index of a function: right hand limit of the function at time t
$\mathbb{Z}$		the set of integers

Bibliography

Alfonsi, A., Fruth, A., and Schied, A. (2010). Optimal execution strategies in limit order books with general shape functions. *Quant. Finance*, *10*(2), 143–157.

Alfonsi, A. and Schied, A. (2009). Optimal execution and absence of price manipulations in limit order book models. SSRN Working Paper 1499209, SSRN.

Almgren, R. and Chriss, N. (2001). Optimal execution of portfolio transactions. *J. Risk*, *3*(2), 5–39.

Almgren, R., Thum, C., Hauptmann, E., and Li, H. (2005). Equity market impact. *Risk*, 21–28. slightly shortened version of 'Direct estimation of equity market impact'.

Asmussen, S. (1987). *Applied Probability and Queues*. Wiley Series in Probability and Mathematical Statistics. Chichester: John Wiley & Sons.

Bachelier, L. (1900). Théorie de la spéculation. *Ann. Sci. École Norm. Sup. (3)*, *17*(21), 21–86.

Bak, P., Paczuski, M., and Shubik, M. (1997). Price variations in a stock market with many agents. *Phys. A*, *246*(3–4), 430–453.

Baxter, M. and Rennie, A. (1998). *Financial Calculus*. Cambridge: Cambridge University Press. first published 1996.

Berglund, N. and Gentz, B. (2006). *Noise-Induced Phenomena in Slow-Fast Dynamical Systems*. London: Springer.

Bertsimas, D. and Lo, A. (1998). Optimal control of execution costs. *J. Financ. Mark.*, *1*(1), 1–50.

Bessembinder, H., Panayides, M., and Venkataraman, K. (2009). Hidden liquidity: An analysis of order exposure strategies in electronic stock markets. *J. Finan. Econ.*, *94*(3), 361–383.

Black, F. and Scholes, M. (1973). The pricing of options and corporate liabilities. *J. Polit. Economy*, *81*(3), 637–654.

Borodin, A. N. and Salminen, P. (1996). *Handbook of Brownian Motion - Facts and Formulae*. Basel: Birkhäuser Verlag.

Bouchaud, J.-P., Gefen, Y., Potters, M., and Wyart, M. (2004). Fluctuations and response in financial markets: the subtle nature of 'random' price changes. *Quant. Finance*, *4*, 176–190.

Bouchaud, J.-P. and Potters, M. (2005). *Theory of financial risk and derivative pricing* (second ed.). Cambridge: Cambridge University Press. reprinted, edition published 2003.

Bovier, A. and Černý, J. (2007). Hydrodynamic limit for the $A + B \to \emptyset$ model. *Markov Process. Related Fields*, *13*(3), 543–564.

Bovier, A., Černý, J., and Hryniv, O. (2006). The opinion game: Stock price evolution from microscopic market modeling. *Int. J. Theoretical Appl. Finance*, *9*(1), 91–111.

Challet, D., Marsili, M., and Zhang, Y.-C. (2004). *Minority Games*. Oxford Finance Series. Oxford: Oxford University Press.

Challet, D. and Zhang, Y.-C. (1997). Emergence of cooperation and organization in an evolutionary game. *Phys. A*, *246*(3-4), 407–418.

Cont, R. (2001). Empirical properties of asset returns: stylized facts and statistical issues. *Quant. Finance*, *1*(2), 223–236.

Cont, R., Stoikov, S., and Talreja, R. (2008). A stochastic model for order book dynamics. Working Paper 1273160, SSRN. accepted for Oper. Res.

Cont, R. and Tankov, P. (2004). *Financial modelling with jump processes*. Chapman & Hall/CRC financial mathematics series. Boca Raton: Chapman & Hall/CRC.

Daniels, M. G., Farmer, J. D., Gillemot, L., Iori, G., and Smith, E. (2003). Quantitative model of price diffusion and market friction based on trading as a mechanistic random process. *Phys. Rev. Lett.*, *90*(10), 108102.

Dischel, R. S. (2002). *Climate Risk and the Weather Market*. London: Risk Books.

Eliezer, D. and Kogan, I. I. (1998). Scaling laws for the market microstructure of the interdealer broker markets. Working Paper 147135, SSRN.

Föllmer, H. and Schied, A. (2004). *Stochastic Finance* (second ed.)., volume 27 of *Studies in Mathematics*. Berlin: de Gruyter.

Frey, S. and Sandas, P. (2009). The impact of iceberg orders in limit order books. CFR Working Paper 09-06, Centre for Financial Research Cologne.

Huberman, G. and Stanzl, W. (2004). Price manipulation and quasi-arbitrage. *Econometrica*, *72*(4), 1247–1275.

Huberman, G. and Stanzl, W. (2005). Optimal liquidity trading. *Rev. Finance*, *9*(2), 165–200.

Hull, J. C. (2000). *Options, Futures and Other Derivatives* (seventh ed.). Prentice Hall series in finance. Upper Saddle River, NJ: Prentice Hall.

Karatzas, I. and Shreve, S. E. (1998). *Brownian Motion and Stochastic Calculus* (second ed.)., volume 113 of *Graduate Texts in Mathematics*. New York: Springer.

Karatzas, I. and Shreve, S. E. (2001). *Methods of Mathematical Finance* (corrected ed.)., volume 39 of *Applications of Mathematics*. New York: Springer. first printing 1998.

LeBaron, B. (2002). Building the Santa Fe artificial stock market. Working paper, Brandeis University, Waltham, MA.

Louis, D. and Müller, P. (2005). *Java 5. Kompendium*. München: Markt+Technik Verlag.

Mandelbrot, B. (1963). The variation of certain speculative prices. *Jour. Bus.*, *36*(4), 394–419.

Mendelson, H. (1982). Market behavior in a clearing house. *Econometrica*, *50*(6), 1505–1524.

Merton, R. C. (1973). Theory of rational option pricing. *Bell J. Econ. Manag. Sci.*, *4*(1), 141–183.

Meyn, S. P. and Tweedie, R. L. (1996). *Markov Chains and Stochastic Stability* (1st ed.)., volume XVI of *Control and Communication Engineering Series*. London: Springer.

Obizhaeva, A. and Wang, J. (2005). Optimal trading strategy and supply/demand dynamics. Working Paper 686168, SSRN. revised and resubmitted, J. Financ. Mark.

Papapantoleon, A. (2008). An introduction to Lévy processes with applications in finance. arXiv:0804.0482v2. Lecture Notes, TU Vienna.

Potters, M. and Bouchaud, J.-P. (2003). More statistical properties of order books and price impact. *Phys. A*, *324*(1–2), 133–140.

Revuz, D. and Yor, M. (1991). *Continuous Martingales and Brownian Motion*, volume 293 of *Grundlehren der mathematischen Wissenschaften*. Berlin: Springer.

Sang, H., Ma, T., and Wang, S. (2001). Hurst exponent analysis of financial time series. *J. Shanghai Univ.*, *5*(4), 269–272. English edition.

Sato, K.-I. (2005). *Lévy processes and infinitely divisible distributions* (english ed.)., volume 68 of *Cambridge studies in advanced mathematics*. Cambridge: Cambridge University Press. reprinted, first published in English 1999.

Schöneborn, T. (2008). *Trade execution in illiquid markets: Optimal stochastic control and multi-agent equilibria*. dissertation, TU Berlin.

Schoutens, W. (2003). *Lévy Processes in Finance*. Wiley series in probability and statistics. West Sussex: John Wiley & Sons.

Simonsen, I. (2003). Measuring anti-correlations in the nordic electricity spot market by wavelets. *Phys. A*, *322*, 597–606.

Smith, E., Farmer, J. D., Gillemot, L., and Krishnamurthy, S. (2003). Statistical theory of the continuous double auction. *Quant. Finance*, *3*(6), 481–514.

Stigler, G. J. (1964). Public regulation of the security markets. *J. Bus.*, *37*(2), 117–142.

Tang, L.-H. and Tian, G.-S. (1999). Reaction-diffusion-branching models of stock price fluctuations. *Phys. A*, *264*(3–4), 543–550.

Voit, J. (2005). *The Statistical Mechanics of Financial Markets* (third ed.). Texts and Monographs in Physics. Heidelberg: Springer.

Weber, P. and Rosenow, B. (2005). Order book approach to price impact. *Quant. Finance*, *5*(4), 357–364.

Weiß, A. (2009). Escaping the Brownian stalkers. *Electron. J. Probab.*, *14*(7), 139–160.

Weiß, A. (2010). Executing large orders in a microscopic market model. arXiv:0904.4131. revised and submitted.

Die VDM Verlagsservicegesellschaft sucht für wissenschaftliche Verlage abgeschlossene und herausragende

Dissertationen, Habilitationen, Diplomarbeiten, Master Theses, Magisterarbeiten usw.

für die kostenlose Publikation als Fachbuch.

Sie verfügen über eine Arbeit, die hohen inhaltlichen und formalen Ansprüchen genügt, und haben Interesse an einer honorarvergüteten Publikation?

Dann senden Sie bitte erste Informationen über sich und Ihre Arbeit per Email an *info@vdm-vsg.de*.

Sie erhalten kurzfristig unser Feedback!

VDM Verlagsservicegesellschaft mbH
Dudweiler Landstr. 99　　　　　　　　Telefon　+49 681 3720 174
D - 66123 Saarbrücken　　　　　　　　Fax　　　+49 681 3720 1749
www.vdm-vsg.de

Die VDM Verlagsservicegesellschaft mbH vertritt

Printed by Books on Demand GmbH, Norderstedt / Germany